Yasuharu Ukai, Editor
**Economic Analysis of Information System Investment
in Banking Industry**

Yasuharu Ukai, Editor

Economic Analysis of Information System Investment in Banking Industry

 Springer

Yasuharu Ukai
Director and Professor, Research Center of Socionetwork Strategies
Kansai University
Suita, Osaka 564-8680, Japan

This English translation is based on the Japanese original, Yasuharu Ukai, editor; Economic Analysis
of Information System Investment in Banking Industry, Published by Taga-shuppan Co., Tokyo
© 2003 Yasuharu Ukai

ISBN 4-431-54637-5 Springer-Verlag Tokyo Berlin Heidelberg New York

Springer is a part of Springer Science+Business Media
springeronline.com
© Springer-Verlag Tokyo 2005
Softcover re-print of the Hardcover 1st edition 2005

Typesetting: Camera-ready by the editor
Printing and binding: Hicom, Japan

Printed on acid-free paper

To Dale W. Jorgenson,
and Shozaburo Fujino

Preface

This is an academic book that explains in reality, examines theoretically, and analyzes statistically information system investment in the banking industry with regard to the process of the information technology revolution. This kind of comprehensive research on the banking industry is the first in the world. It could be seen as an application study for Japanese financial deregulation after 1997. However, our project, the Workshop of Information System Investment, is a theoretical research venture, consisting originally, when it began in 1994, of economists and computer scientists. It aimed to measure the effect of computer hardware and software on the modern economy, based on the microdata of each firm, and to extend the frontiers of economic science. It was, coincidentally, the time when this project began full-scale operation, in July 1997, that the voluntary closure of Yamaichi Securities was decided. The failure of the Hokkaido Takushoku Bank was disclosed in November of the same year, and the breakdown, temporary nationalization, buying out, and mergers of several banks succeeded one another. Our research therefore suddenly got into the social spotlight on the application stage.

Part I is the first history and strategic guidelines of information systems in the banking industry. Part II summarizes the economic analyses of information system investment in the United States, Europe, and Japan. These parts are foundations for the statistical analyses in Part III. Part III categorizes information system investments in the banking industry into three types and analyzed them. The treatment of system development and personal expence in economics is quite different from it in accounting theory. First, the investments in computer hardware, peripheral equipment, computer software, knowledge asset concerning system development, and computer related human resource are defined as Information System Investments I. The accumulated value of this kind of investments are defined as Information System Assets I. Second, the investments in computer hardware, peripheral equipment, computer software, and knowledge asset concerning system development are defined as Information System Investments II. The accumulated value of those are defined as Information System Assets II. These types of investments and assets will be explained in Chapters 5-9. Third, the investments on the information system including strategic planing, product development, and customer service are defined as Information System Investments III. The accumulated value of those are defined as Information System Assets III. This definition will be critical for the analysis in Chapter 5. Information System III is near concept of strategic information system (SIS) in the USA. Finally, computer hardware and software investment, and its asset will be analyzed in Chapter 9.

The results of our analyses contradict the prevailing pessimistic expectations for the financial technical revolution in Japan. First of all, based on replies to our questionnaires, the positive effect of information system investment I and II on the market value of a bank is considerable. Its numerical value is not greatly different from that in the USA manufacturing industry during the boom. Second, we also found out that the computer software had a much stronger effect on firm's performance than the hardware.

This conclusion will shock economists performing econometric analysis on the basis of disclosed data on the industry and macroeconomic levels. We are preparing ourselves for strong criticism and are looking forward to constructive discourse.

As this research advanced further, we faced the stark fact that individual statistics of computer investment by Japanese firms had not been completely disclosed. The computer-related items in public financial statements were scattered among computer hardware and software items, and the principles of accounting procedure were not unified until April 1999. Therefore, we are deeply grateful to the information system specialists at each bank for their academic advices. We will withhold their identities, because the material they provided must be kept confidential. However, we want to mention Mr. Yoshiaki Sugita (former director of the system planning department at Fuji Bank), who was one of the leaders in the field of information technology in the banking industry.

We also wish to gratefully acknowledge the following persons: Prof. Dale W. Jorgenson (Harvard University), Dr. Daniel E. Sichel (Federal Reserve Board), Prof. Erik Brynjolfsson (Massachusetts Institute of Technology), Prof. Marshall Van Alstyne (Boston University), Dr. Charles King, Prof. Akihiko Shinozaki (Kyushu University), Prof. Kazuyuki Suda (Waseda University), Prof. Mitsuru Iwamura (Waseda University), Prof. Kenji Shiba (Kansai University), Ms. Lilia Shiba, Prof. Yoichi Iwasa (Kansai University), Prof. Koichi Takeda (Hosei University), Prof. Hiroaki Aoki (Hannan University), Prof. Hideki Nishimoto (Ryukoku University), Mr. Toshitaka Hagiwara (Financial Accounting Standards Foundation), Mr. Hajime Mizuno (Bank of Japan), Dr. Kimiaki Aonuma (Bank of Tokyo-Mitsibishi), Mr. Kiyoshi Maenaka (Resona Bank), Mr. Toru Moroga (Resona Bank), Mr. Shigeo Kiuchi (Resona Bank), Mr. Yoshiki Oshida (Resona Bank), Mr. Junjiro Kunihiro (Resona Bank), Mr. Sunao Fujimoto (Resona Bank), Mr. Kenichi Tanai (Resona Research Institute), Mr. Tokuya Sugamiya (Hitachi Corporation), Mr. Toshikazu Yonekura

(Hitachi Corporation), Mr. Motohiko Ohnishi (The Center for Financial Industry Information Systems). In particular, I deeply appreciate the academic advice received from Prof. Jorgenson since 1981 when I first set foot in Boston as a Fulbright Fellow at Harvard University.

This research project has received six grants from Japan Society for the Promotion of Science, as follows. Grants-in-Aid for Scientific Research, C(1) 1997-2000, "Economic Analyses of Information System Investment in the Banking Industry" (project number 09630061); B(2) 1999-2002, "The Development of an Economic Evaluation Method of Information System Investment in the Financial Industry" (project number 11553001); B(2) 2001-2004, "The Micro Data Analyses of Information System Investment in IT-related Industries" (project number 13430019); Ministry of Education, Culture, Sports, Science and Technology, Academic Frontier Promoting Project, 2002-2006, "Policy Studies of Network Strategies as a Social Infrastructure"; Grant-in-Aid for Publication of Scientific Research Results, 2002 and 2004 "Economic Analysis of Information System Investment in Banking Industry." (project numbers 145261 and 165256)

Our research raised many questions that need to be addressed in microeconomics, financial systems, and information technology analysis. We hope that it becomes a milestone in the new academic field of information system investment.

Yasuharu Ukai
Director, Research Center of Socionetwork Strategies, Kansai University
Senriyama, Osaka, Japan,
December 2004

Contents

Part I Past and Present of Information System Investment in the Banking Industry

1   Past and Present of Information Systems in Banks   3
   1.1   Chronological Changes in Information Systems in Banking   3
       1.1.1   Off-Line Systems   3
       1.1.2   Progress of On-Line Systems   5
       1.1.3   Networks around the Banking Industry 16
   1.2   Accidents That Affected On-Line Systems and the Lessons
         Learned 17
       1.2.1   Setagaya Cable Fire 17
       1.2.2   The 1992 London Explosion 20
       1.2.3   Great Hanshin Earthquake 21
       1.2.4   The Year 2000 (Y2K) Problem 23
   1.3   Roles of the System Department 24
       1.3.1   History of the System Department 25
       1.3.2   The Changing Roles of the System Department 27

2   Information System Strategy of Nationwide Banks 29
   2.1   Current Banks and Systems 29
   2.2   Management Subjects of the Banking System 30
       2.2.1   Countermeasures for Changing Conditions 30
       2.2.2   Strategy for Competitive Advantage 30
       2.2.3   Adjustment to New Technology 32
       2.2.4   Strategic Support System 32
       2.2.5   Improvement of Customer Services 32
       2.2.6   Risk Management 34
       2.2.7   Maximizing Business Efficiency 34
       2.2.8   System Management 35
   2.3   Revolution of Delivery Channels 35
       2.3.1   Changes to the Point of Business in the Bank 36

    2.3.2 Network Banking Strategy........................ 39
2.4 Creation of Mega Banks and System Problems 42
2.5 Issues 44
    2.5.1 Technical Issues 44
    2.5.2 System Management 47

Part II Review of Information System Analyses

3 **Limit of Aggregate Level Analysis of Information System
   Investment** 55
   3.1 Introduction 55
   3.2 Productivity Paradox 57
   3.3 Studies of the Whole Economy 58
       3.3.1 Growth Accounting 59
       3.3.2 Macro Production Function and Cost Function 62
       3.3.3 Effect on Consumer-Surplus 63
   3.4 Industry-Level Studies 64
   3.5 IT Effect on Employment for the Whole Economy 67
   3.6 Summary... 68

4 **Firm-Level Analysis of Information Systems Investment** ... 71
   4.1 Introduction 71
   4.2 Studies on Productivity 72
   4.3 Contribution of IT Capital to Firm Value.................. 78
   4.4 Analysis of Organization Effect and Management Strategy
        Effect 81
   4.5 Effect on Working Conditions 84
   4.6 Summary... 86

Part III Positive Analyses of Information System Investment in the Banking Industry

5 **Outlook and Study Process of Questionnaires**.............. 91
   5.1 Background and Study Process of Questionnaire Surveys 91
   5.2 The Concept of Information System Investment 99
   5.3 Accounting Problems in Information System Investment...... 101
   5.4 Lessons from Questionnaire Surveys 104

6 **Disclosure and Circumstances Concerning Information
   System Assets** 107
   6.1 Introduction 107
   6.2 Accounting Procedure of Information System Assets 108
       6.2.1 Accounting Procedure (1): Computer Software Assets .. 108

6.2.2   Accounting Procedures (2): Computer Equipment 109
6.2.3   Accounting Procedure of Research and Development
       Costs 110
6.3   Computer Software as an Asset 110
6.4   Types of Cooperate Disclosure 111
   6.4.1   Legal and Voluntary Disclosure 111
   6.4.2   Necessity of Disclosure 112
6.5   Information System Assets in Japanese Banks.............. 113
   6.5.1   Information System Assets 116
   6.5.2   Computer Software 118
   6.5.3   Computer Equipment 120
   6.5.4   Planned Budget for Information System 122
6.6   Classification of Banks by Information System Assets 123
6.7   Summary.. 125

7   Cross-Section Analysis of Information System Investment . 127
7.1   Introduction 127
   7.1.1   Definition of Information System Investment and
       Questionnaire Items 128
   7.1.2   Financial Index and Total Employee Number in
       Answering Banks.................................. 129
7.2   Various Statistics of Questionnaires 130
7.3   Results of Regression Analysis 143
   7.3.1   Information System Development Cost and Loan and
       Bills Discounted.................................. 144
   7.3.2   Information System Development Cost and Total Assets 144
   7.3.3   Information System Development Costs and Number
       of Personnel 145
7.4   Management Strategy and Investment Activity 146
7.5   Conclusions.. 148

**8   Analysis of Information System Investment Using
Questionnaire Data** 149
8.1   Introduction 149
8.2   Estimated Model and Data Set........................... 151
   8.2.1   The Market Value Model: Brynjolfsson and Yang
       Type Model 151
   8.2.2   Data Sources and Construction 153
8.3   Estimated Results 157
8.4   Conclusion 162

**9   Analysis of Information System Investment Using Public
Data**... 165
9.1   Introduction 165
9.2   Analytical Tools and Data Set 166

       9.2.1   Formation: Model for Panel Data.................... 167

       9.2.2   Data Sources and Construction 167

       9.2.3   Unbalanced Panel Data and Balanced Panel Data 172

  9.3   Estimation 173

       9.3.1   Estimation (1) Unbalanced Panel Data Analysis....... 174

       9.3.2   Estimation (2) Balanced Panel Data Analysis 180

  9.4   Concluding Remarks: Future Directions and Opportunities
for Research 182

10 Conclusion 187

Part IV Appendix

A   Information System Investment Questionnaires 1995-1997 . 193

B   Documents about Accounting Standards 207

C   Mathematical Appendix 217

References 221

Index 231

List of Figures

1.1 Enlargement of business and shifting branches to the new
     system 7
1.2 General scheme of the shift to the second-generation on-line
     system 11
1.3 General figures of banking network systems 13
1.4 The third-generation on-line system and the hub-and-spoke
     type system 14

2.1 Sources of risk surrounding the computer system 35
2.2 Hub-and-spoke type branch networks 37

6.1 Changes in the number of banks in Japan based on charging
     computer software 114
6.2 Proportion of banks based on the frequency of writing
     computer software in the financial statements more than one
     times in the 1993-1999 period. 115
6.3 Scatter diagram between computer software assets and
     computer equipment for Japanese banks 117

9.1 Share of bad debt in loan and bills discounted in Japanese
     banks for 1993 and the 1997-1999 period 170
9.2 Number of Japanese banks using on-line systems at the time ... 171

List of Figures

List of Tables

1.1   Progress of on-line systems in the banking industry   6
1.2   Lessons from Setagaya cable fire   18
1.3   Lessons from the 1992 London explosion   21
1.4   Lessons from the great Hanshin earthquake   22
1.5   Countermeasures to the Y2K problem   24

2.1   A list of main management problems, main themes, and items
      facing information systems   31
2.2   Main services of firm banking and home banking   33
2.3   ATMs in main convenience stores   39

3.1   Studies of IT investment effect on the whole economy   58
3.2   Industry-level studies on economic effect of IT investment   64

4.1   Firm-level studies on economic effects of IT investment   72

6.1   Number of Japanese banks in the 1993-1999 period 114
6.2   Statistics of information system assets in Japanese banks for
      each year during 1993-1999 116
6.3   Statistics of computer software in the 1993-1999 period 119
6.4   Statistics of computer equipment in the 1993-1999 period: I 120
6.5   Statistics of computer equipment in the 1993-1999 period: II . . . 121
6.6   Statistics of the planned budgets concerning information
      systems in the 1993-1999 period 123

7.1   Statistics from answers to the first questionnaire 131
7.2   Statistics from answers to the second questionnaire 132
7.3   Statistics from answers to the third questionnaire 133

8.1   Estimated result using unbalanced panel data (Information
      System I) 158

8.2   Estimated result using unbalanced panel data (Information
System II) 158
8.3   Estimates of regional banks using balanced panel data
(Information System I) 160
8.4   Estimates of regional banks using balanced panel data
(Information System II) 160
8.5   Unbalanced panel estimates with the outsourcing variable
(Information System I) 161
8.6   Unbalanced panel estimates with the outsourcing variable
(Information System II) 162

9.1   Number of bank disclosed information concerning computer
software assets in financial statements 172
9.2   Estimated result using unbalanced panel data I 174
9.3   Estimated result using unbalanced panel data II 175
9.4   Estimated result using unbalanced panel data III (nationwide
banks) 177
9.5   Estimated result using unbalanced panel data IV (nationwide
banks) 177
9.6   Estimated result using unbalanced panel data V (regional banks) 178
9.7   Estimated result using unbalanced panel data VI (regional
banks) 179
9.8   Estimated result using balanced panel data I 181
9.9   Estimated result using balanced panel data II 181

Authors

Yasuharu Ukai Director and Professor, Research Center of Socionetwork Strategies, Kansai University
   3-3-35 Yamate-cho, Suita, Osaka, Japan 564-8680
   ukai@rcss.kansai-u.ac.jp

Shinji Watanabe Assistant Professor, College of Integrated Arts and Sciences, Osaka Prefecture University
   1-1 Gakuen-cho, Sakai, Osaka, Japan 599-8531
   shinji@hs.cias.osakafu-u.ac.jp

Hisao Nagaoka Director, Osaka Securities Finance Co., Ltd. and Associate Fellow, Research Center of Socionetwork Strategies, Kansai University
   3-3-35 Yamate-cho, Suita, Osaka, Japan 564-8680
   nagaoka@rcss.kansai-u.ac.jp

Toshihiko Takemura Research Assistant, Research Center of Socionetwork Strategies, Kansai University
   3-3-35 Yamate-cho, Suita, Osaka, Japan 564-8680
   takemura@rcss.kansai-u.ac.jp

Part I

Past and Present of Information System Investment in the Banking Industry

1

Past and Present of Information Systems in Banks

H. Nagaoka, Y. Ukai, and T. Takemura

1.1 Chronological Changes in Information Systems in Banking

1.1.1 Off-Line Systems

In the latter half of the 1950s, the banking business became popular as the Japanese economy grew at a remarkable rate. The banks started services such as the provision for ordinary deposits to be payable in all branches, personal checks, and account transfers. Therefore, as the volume of business increased, the banks required a larger labor force to implement operations by conventional methods, including counting cash by hand, posting up accounts in ledgers, and use of the soroban, the Japanese classical abacus. Much of the typical banking business was suitable for machine-based processing and thus the bank management realized that rationalization and automation of many aspects of the business were important.

As a result, the punched card system (PCS)[1] was installed and usually used for the statistical calculation. This technical process was the beginning of machine-based processing. Some banks also used a batch-processing system[2], and some banks employed many key-punchers[3] for the data entry tasks.

---

[1] PCS was used whereby transaction data was punched onto cards and could be read automatically. PCS was used up to about 1960 and was then replaced by the computer. See Daiwa Bank History Committee (1979) p147.

[2] Batch-processing is a type of data processing in which large amounts of data may be gathered for many transactions before the transactions are actually processed as a group. Today, on-line systems and real-time processing has spread all over the world and have largely replaced batch-processing.

[3] Key-punchers are people who punch the data on cards and paper tapes with a card punch machine and paper-tape punch machine.

These processing systems were applied to calculation of statistical data, payroll details, discount charges, posting up of accounts for commercial papers, daily trial balances, and so on. Ordinary deposits were also conducted using the off-line system[4]. In addition, teletypewriters[5] were installed in all branch offices so that fund remittances and correspondence could be sent and received among the branch offices. These operational devices were the starting point for the development of on-line domestic exchange.

With the increase of clearance volumes, mechanized operation of business became an acute task. The use of magnetic ink character recognition (MICR) was adopted for mechanization of clearing. Standards for printing letters were established by the Federation of Bankers Associations of Japan[6] in April 1965, and standardized horizontally written bank checks were brought into use in November 1968. This standardization contributed to rapid progress in processing checks and notes.[7]

Meanwhile, accounting machines had already been installed to deal with the business of ordinary deposits and current accounts in all branch offices. However, in some branch offices, the calculation of interest for ordinary deposits was performed manually by staff.

With the setting up of teller machines, calculators for bank notes and coins, many kinds of unit machines were utilized in rationalizing banking business. Nevertheless, the functions of these machines were different from those currently in use. The former could count notes and coins but could not bind a batch of notes or coil coins. These tasks still required the use of manual labor.

The collection center for bills and checks was established as a proxy for all branch offices. The establishment of the collection center contributed to reducing the workload in branch offices and cutting postal and telephone costs. As a result, banks improved a correspondence policy thoroughly.[8]

Business efficiency improved as a result of such mechanization and rationalization. On the other hand, the business of posting up accounts by pen and ink still remained. At that time, the bank adopted the ballpoint pen as an official writing tool. Typists made up applications for financing or loans by

---

[4]This is a processing measure that does not link the computer and terminal machines by communication lines. Before computer processing, it is necessary to obtain data through manual processing.

[5]The teletypewriter is a kind of teleprinter. If codes are sent by the teletypewriter, they are sent to receivers through communication lines. The received codes are then automatically translated to characters and then are printed. See Daiwa Bank History Committee (1988) pp150-151.

[6]The Japanese Bankers Association was renamed from the Federation of Bankers Associations of Japan since April 1999.

[7]See Tsuchiya, Harada, and Tomaru (1983) p170, and p173.

[8]It is a kind of clearing system that can settle financial transactions with other banks without the transportation of cash. See Daiwa Bank History Committee (1988) pp150-151.

Japanese-character typewriter and prepared certified checks and certificates with the check writer. In such circumstances, it was not always possible to systematically promote mechanization and rationalization. Rather, the banks gradually accepted mechanization and rationalization as it became possible. In summary, banks have steadily promoted the mechanization of business processes in a step-by-step fashion, which in turn has led to further systematic changes.

1.1.2 Progress of On-Line Systems

An on-line system is generally defined as a mainframe computer connected with many terminal units. The central mainframe computer controls all of the terminals, and data are exchanged on a real-time basis.

The on-line systems of banks have dramatically evolved through three waves of information technology innovation, which have occurred in every decade since 1965. The first wave of change is generally recognized as the first-generation on-line systems. This system operated in the 1965-1975 period, and was mainly used for data processing within the individual bank network system as an alternative to the off-line system. Up to that time, banks had used the off-line system for batch processing.

The second wave of information technology innovation, known as the second-generation on-line system, operated in the 1975-1985 period. This expanded the network system from an isolated system to one that interacted with those of other banks and customers.

As more bank networks expanded, the business conducted on them became more complex. This was the third wave of information technology in the 1985-1993 period called the third-generation on-line system.

At present, the on-line system is in the post-third-generation phase (from about 1993), with the various forms of advanced and complex information technology in use being extensions of the third-generation on-line system rather than a replacement of it. The development of the banking information system in Japan is shown in Table 1.1, along with the aims and characteristics of the system.

The First-Generation On-Line System

As banking services gradually became more popular, banks faced the limitations of off-line systems in dealing with increasing volumes of business and in improving customer services. It soon became essential for banks to reduce the size of their labor forces and to introduce on-line systems. In 1965, the Mitsui Bank (now a part of Sumitomo Mitsui Banking Corporation) was the first Japanese bank to start an on-line system for ordinary deposits. Afterward, every nationwide bank in Japan followed this lead and implemented on-line systems. Meanwhile, some banks in the United States had adopted such systems before the Japanese banks. In particular, some medium or small-sized

Table 1.1. Progress of on-line systems in the banking industry

	First-generation on-line system	Second-generation on-line system	Third-generation on-line system	Post-third-generation on-line system
Purposes	Reduction of labor force, and rationalization of office work	Rationalization and improvement of customer service	Deregulation of monetary system, and improvement of administrative information	Improvement of new product, delivery channel and IT strategy, total risk management, fulfillment of customer network, synthesis of system, and upgrade of clearing system
Features	Dealing with single subject (Systemizing ledger in on-line system, and centralization of automatic transfer center)	Connective processing of main subjects, fulfillment of consolidated account, and interbank cooperation of on-line CD	Re-establishment of higher account system, and establishment of several subsystems and organic combination	Many independent subsystems, spread of broadband and ubiquitous banking, promotion of IC card, spread of open system, and improvement of security measures
Networks	From intrabank network to interbank network	From interbank network to interindustry network	From interindustry network to PC network.	From PC network to Internet
Range of control	Intrabank (domestic) Start of EB Interbank		Intrabank (international) Enlargement of EB Variety of interbank	Open network
Investment amount	15-20 billion dollars per bank	25-35 billion dollars per bank	50-80 billion dollars per bank	
Changes to form of the organization	Reduction of 1000-2000 employees by broadening section responsibility	Reduction of 2000-3000 employees by streamlining	Reduction of 1000-2500 persons by further streamlining	
	Developing age of system department		Expansion and subdivision of system department	Rationalization of system department and outsourcing

CD (cash dispenser), EB (electronic banking), PC (personal computer), IC (integrated circuit),

Reproduced from the Center for Financial Industry Information Systems (2000a) p44 with modifications banks in the USA adopted on-line systems within the small regions in which they operated.[9]

It could be said that the first-generation on-line systems adopted by Japanese nationwide banks were epoch-making, because the on-line systems could cope with a much larger volume of data more easily than the old system and had the widest and largest networks that stretched all over Japan.

Along with the on-line system, developments were gradually introduced, starting from, say, ordinary deposits and domestic exchange, to current account, deposit at call, term deposit (time deposit), loans, foreign exchange, and daily trial balance. Finally, a totally integrated on-line system was obtained when all the business units were systematized. Fig. 1.1 shows how each business unit was integrated into the total on-line system.

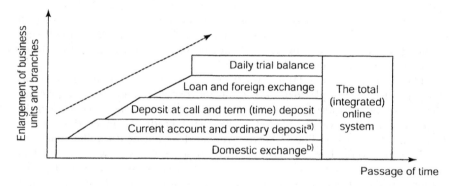

a) Current account and ordinary deposit were implemented in the new system in groups of 5-10 branches at a time

b) Domestic exchange and diary trial balance were renewed in all branch offices simultaneously for the continuity of business

Fig. 1.1. Enlargement of business and shifting branches to the new system

Generally speaking, ordinary deposits were gradually shifted into the new system in groups of 5-10 branch offices. At the same time, domestic exchange operations were renewed at all branch offices. The activities of current account, deposit at call, term deposit (time deposit), loans, and foreign exchange were changed gradually.

As a result of rationalization of managing accounts, posting up of accounts, and calculation of interest, banks reduced work forces by thousands. Hence, the introduction of the first-generation on-line system led to rationalization of office work and a reduction of labor forces. Even after shifting to the totally integrated on-line system, the office work operations in branch offices remained

---

[9]Dealing with checks and bills was a major issue for large banks in the USA.

vertically in each section such as ordinary deposits section, current accounts section, domestic exchange section, and so on.[10]

The on-line system enabled the ordinary deposits to be made at all branch offices, concentrated on the clearing of checks and notes, and automated fund transfers, which were not possible with the off-line system. Furthermore, the on-line system contributed to strengthening of business management by including systematic checking on bookkeeping.

In September 1971, the public telecommunications law was revised and the first stage of deregulation of telecommunication lines was realized.[11] The Domestic Funds Transfer System, called the Zengin system[12], was established by connecting domestic banks after this deregulation. In April 1973, this system began to operate at full scale. Before the introduction of the Zengin system, exchange dealing was made on the basis of "correspondent arrangement," an independent agreement between corresponding branch offices. By contrast, the Zengin system was able to transfer the domestic remittances among all branch offices in banks as one integrated system. The system dramatically improved the speed of the telegraphic money order to other banks and reformed the domestic remittance in the banking industry.

The cash dispenser (CD) was put to use at around this time. The CD was first installed on the platform of the off-line system. Therefore, the primary CD occasionally required much time to confirm the customer's account on the basis of the off-line system. These CDs were then gradually replaced by on-line CDs and afterward on-line CDs were replaced by automatic teller machines (ATMs) that were able to receive or draw on accounts. After CDs were introduced, customers were able to withdraw money themselves and without seeing cashiers at the teller window. Not only did CDs gain favor with customers, but also they presented banks with a sizeable benefit from labor cost reduction. In the early stages, bank leaders felt quite uneasy about concerns from their managers. These included whether CDs could service customers adequately, and whether sales opportunities at the teller window might decline because of decreased face-to-face transactions. In addition, some customers felt that they could not obtain incidental information by using a CD or felt uncomfortable about CD operation.

At that time, an integrated system for ordinary accounts and term deposits (time deposits) called sogo koza contributed to the rationalization of office work in branches. This was conducted in around August 1972 and gradually combined other banking products. The system was a pioneer of systematic products, and these were rapidly popularized as they met favor with customer needs.[13]

---

[10]See Yamada, and Sekiguchi (1989) pp6-9.

[11]Earlier, the Nippon Telegraph and Telephone Public Corporation (now NTT) had the monopolistic right to use domestic communication lines.

[12]The Zengin System operating on a national scale and providing information and settlement facilities for the electronic funds transfer.

[13]See Iwasa (1990) p113.

The first-generation on-line system became the foundation of the system that has evolved to the post-third-generation on-line system. The initiation of the first-generation system was a landmark event. The totally integrated on-line system brought about the reduction of office work concerning the management of accounts and the calculation of interest and promoted a more lateral organization of office work within branch sections. However, office work in branches still remained vertically integrated according to conventional banking sections. Furthermore, there were various problems associated with the first-generation on-line system related to customer services and the entry of data into the information system.

The Second-Generation On-Line System

The first-generation on-line system developed each account step by step as shown in Fig. 1.1. The system was able to rationalize office work and reduce the size of the labor force. However, there were inconveniences related to the retrieval of all accounts held by individual customers, and it became increasingly important to develop system capability for over the retrieval of all accounts according to customer name and transaction type. They were management challenges related to data entry and administration of the information system, in addition to rationalization of office work, as in the age of the first-generation on-line system.

Advances in information technology from upgrades to the operating system (OS), high-speed processing of the mainframe computer, large memory capacity made remarkable progress. Challenges in the system were solved gradually, and many system troubles that were connected with the overall system-down at the level of the first-generation system persisted still. On the other hand, in the second-generation on-line system, the system security was dramatically improved and problems which minor troubles exercised influence on other systems were resolved.

With the progress of information technology, this system was capable of dealing with a wider range of connective treatments than the first-generation on-line system, and showed improved efficiency in the use of the customer information file (CIF). The system improved transaction connections between subjects, such that a transfer from one account to another could be performed by only one operation. Therefore, by dealing with debit notes on debt and credit transactions, receipt notes were made in the computer data file as system notes. This process reduced the number of notes, and, in doing so, rationalized office work and helped to reduce the labor force. The system adopted a ledger file system for each customer and was then able to gather names over all accounts for each customer.[14]

Each bank not only increased the number of CDs and ATMs in the internal network, but also made progress in cooperation with other banking organizations concerning on-line CDs and ATMs. Products of electronic banking

---

[14]See Iwasa (1990) p113.

(EB) were improved and they contributed to strengthen customer services so that customers utilized EB more and more.[15] Therefore, many banks installed what is known as a front-end system[16] to decrease the burden of all transactions occurring via the host computer. Essentially, the front-end system was aimed at the efficient processing of transactions despite the rapid increase of data volume. It is likely that the system was also effective in enhancing security. Banks experienced various problems and accidents, such as the Setagaya cable fire (see Subsect. 1.2.1), at about the time of the first-generation on-line system. Therefore, plans were made to strengthen security, and included the decentralization of the computer center and duplication of communication lines.

At this time, the office work systems in branch offices changed from vertically integrated organizations to more laterally organized operations, such as the teller section, back office section, etc. For example, the teller section at the window was divided between the high counter and the low counter. The former dealt with comparatively simple transactions and covered the range of payment/withdrawal of ordinary deposits, domestic exchange, payment of public charges, etc. The low counter was used to counsel customers when opening accounts and undertaking new business, or simply for the provision of advice. By changing the office work system, specific sections could avoid confusion at the customer window because of the cooperation of members within a large section.

Consequently, the operational restructuring shortened waiting times and placed more emphasis on customer service, but also reduced labor force numbers much more so than the first-generation on-line system. On average, each bank reduced the number of staffs by about 2000-3000. Another reason for this dramatic reduction in staff numbers was realized by promoting the concentration of back office works.[17]

The second-generation on-line system was able to deal with additional subjects that the first-generation on-line system could not. In fact, all subjects of banking business were switched over from the former system to the new one all at once. This approach was different from the transfer of each single subject in the first-generation on-line system. In this transformation period, both the former system and the new one were run in parallel. It was expected that the system transfer would be conducted over a short period with a view to reducing costs. As shown in Fig. 1.2, all subjects were transferred to the new system in groups of about ten branches. Therefore, the number of branches using the new system was gradually increased in a stepwise fashion. Upon

---

[15]Refer to Chap. 2 for a discussion of EB products.

[16]Front-end systems were introduced in the period from the latter half of the first-generation to the second-generation on-line system. See Daiwa Bank History Committee (1988) p360.

[17]See Yamada, and Sekiguchi (1989) p31.

the completion of the second-generation on-line system, the first-generation on-line system was discarded.

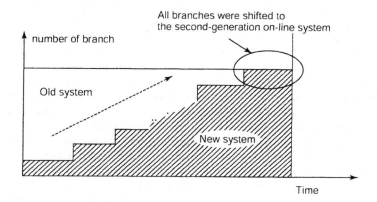

All aspects of the banking business were shifted to the new system in groups of 5-10 branches at a time

Fig. 1.2. General scheme of the shift to the second-generation on-line system

The Third-Generation On-Line System

In the mid-1980s, banks increasingly developed additional and complicated functions of information systems as deregulation and expansion of banking business continued. As this process unfolded, it became clear that the second-generation on-line system had several functional limitations and restrictions. As banks soon reached the limitations of systems in dealing with securities business, international business, and information, the age of the new integrated on-line system dawned. This required the functional combination of subsystems and the main account system. Therefore, banks began to develop the third-generation on-line system to address various problems.

The aim of the third-generation on-line system was to provide flexible management in adapting to the deregulation of banking business, internationalization, various needs of customers, improvement of window services, and securities. At this time, the role of the information system expanded to not only deal with office work, but also plan management strategies in the bank. It is said that the third-generation on-line system was unprecedented from the viewpoint of high-speed processing of applications and the transaction of various business functions. Of note, it is pointed out that the extended availability of the service (7 days/24-h) is one of the features of the system. This function was not perceived as being immediately necessary, but was designed from the beginning because it was expected that customers would require the function in the near future.

The third-generation on-line system realized the system-maintenance during on-line system operated, and the delayed-processing. (batch-processing in the daytime) In the second-generation on-line system, batch-processing during the night time after closing the on-line system generated management reports, performed system maintenance, installed new computer software, and offered the data to other subsystems, etc. However, this approach began to suffer when batch-processing after closing time was unable to be completed in the available time, particularly when the operating time of the on-line system was extended. Therefore, it was necessary to develop a system capable of maintaining on-line operation while other processing functions were performed. At the same time, the method of updating data files during on-line operation for integrating data toward other subsystems was adopted gradually. It was planned that the security policy would be strengthened by hot standby[18], etc.

Ordinary office work became more efficient at the teller window because of the reduced need for consultation with other staff. Transactions that required a decision or approval from the manager could be obtained on the terminal display by the transmission of data from the clerk to the office manager. Therefore, clerks could deal with every transaction without leaving the teller window. The fact led to reduced waiting times for customers, while clerks could easily seek information regarding a customer's business on the terminal display. This system also contributed to improving sales at the teller window. With the incorporation of the above-mentioned features, it is said that the third-generation on-line system became a truly integrated account system by the host computer.

With the wide prevalence of ATMs, these machines were given multifarious functions in the third-generation on-line system. ATMs had already replaced most CDs, and about 70%-80% of customers entering branches did so to use an ATM. If 1000 customers entered a branch in a day, about 700-800 of them were able to satisfy their needs in the ATM at "quick lobby." This is because the bank and the computer maker designed the ATM to provide convenience and a wide range of banking functions. In the early stages, the installation site for CD/ATMs was called the "cash corner" or "quick corner," although this was nothing more than a few CD/ATMs in a corner of the branch office. Therefore, banks regarded CD/ATMs as a supplement to cashiers, with the primary functions of reducing the burden on cashiers, shortening waiting times, and cutting down the cost of office work. In the second stage of on-line development, banks installed a number of ATMs on the main floor of branch offices and called the area the "quick lobby." In branch offices that received large numbers of customers, banks installed many ATMs on the first floor wholly. In such branch offices, it became increasingly common for cashiers and

---

[18]Hot standby is a setting of the system that allows immediate switching to the backup computer when accidents occur. See Katagata (1989) p79.

managers to work on the second floor of the branch building. Banks gradually changed the role of ATMs from that of a support service to the leading role, as reflected by the change in terminology from "corner" to "lobby."

The CD/ATM on-line systems of nationwide banks were connected to each other by a system known as the Bank Cash Service (BANCS). In addition, each type of banking organization was connected with the Multi Integrated Cash Service (MICS). This meant that all CD/ATM on-line systems became totally connected, and were therefore quite complicated. Fig. 1.3 shows the general structure of the banking networks.

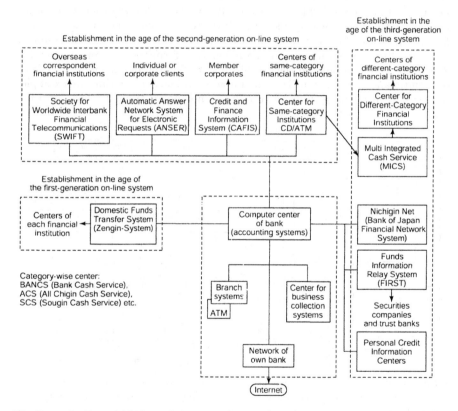

The Center for Financial Industry Information Systems (2000a) p45, and (2004a) p6 with modifications

Fig. 1.3. General figures of banking network systems

Banks in Japan promoted planning and development around the time of the third-generation on-line system and showed remarkable results during this time. After the collapse of the bubble economy, banking results dropped rapidly. However, banks had already invested heavily in the on-line system; the investment in the third-generation on-line system was larger than the first- and the second-generation on-line systems. It is now said that the third-generation on-line system was developed with good timing from the viewpoint of the severe financial situation today.

Post-Third-Generation On-Line System

In the time following the development of the third-generation on-line system, the system department was required to deal with the complexity of customer needs and to develop new products quickly and efficiently. In order to adapt to this new situation, it was necessary to invest enormously and develop the information system for the long term. After the bursting of the economic bubble of 1989-1990, it was difficult for the banks to reform the system completely. Therefore, banks promoted the establishment of a system with improved ease of maintenance and high utilization of functions. During this time, an example of this reform of the system was the "hub-and-spoke" type, as shown in Fig. 1.4.

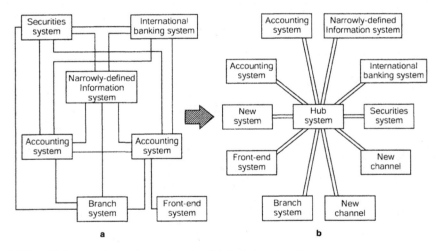

(a) The third-generation on-line system and (b) the hub-and-spoke type system

M Ito (1999) p21 and the Center for Financial Industry Information Systems (2000a) p287 with modifications

Fig. 1.4. The third-generation on-line system and the hub-and-spoke type system

This system was able to integrate several subsystems using middleware such as "message broker."[19] A system of the hub-and-spoke type has advantages as follows:

---

[19]See M Ito (1999) p21, and Togashi (2000) pp57-63.

1. It is capable of adding new functions during the system operation without influencing it.
2. It makes the integration and separation of the system easy, and realizes business cooperation and outsourcing efficiently.
3. It restricts the influence of system trouble to a narrow area and improves the maintenance of system.

It is thought that these functions exercised power over various cases of integration, and joint utilization as mergers and acquisitions of banks proceeded. The adjustment of these fundamental systems became an increasingly important factor of management strategy. Moreover, in the information system strategy for retailing, banks promoted the expansion of ATM networks and the variety of delivery channels. These facts strongly influenced managers' duties in branch offices.

During the first- and second-generation on-line systems, the administrative range was within the network of leased lines and administrative management within branches or within the bank was sufficient. However, the range of administrative management was enlarged enormously with the expansion of ATM cooperation and the increase of transactions in the open network.[20] Therefore, managers in every branch office were required to understand the mechanism of networks and systems, obtain administrative techniques and skills, and deal with customer needs in cooperation with the "customer center," and so on.

The issue of the so-called Y2K problem also forced a heavy burden on the bank management. This is described in detail in Subsect. 1.2.4. While banks dealt with the Y2K problem, they postponed further development of the on-line system. After safely dealing with the Y2K issue, banks then resumed active promotion of investment in the information system. During this time, a large number of themes to develop systems for network banking and risk management emerged. As several large nationwide banks undertook mergers and acquisitions, the integration of each bank system became very important. Chap. 2 discusses these problems further. If one bank can unify its own system with that of another, the need to invest in central machines, terminal machines, and computer software is met at reduced cost.

It is clear that after unification of systems, considerable savings will be made in the long term. System unification also involves a unity of business procedures, and, as a result, training and transition to unification has become a major project for all parts of the banks concerned. Despite such processes, the main accounting systems of the banks are still operated by large-scale mainframe computers. In the process of system unification, such concepts

---

[20]The range of business products was diversified not only by EB facilities and networks in the banking industry, but also by the diversification of international networks such as global networks, international cash management services (CMS), international cards, and so on.

would naturally be applied. It is generally assumed that if a bank were to develop a new accounting information system, it would need a long development time and massive investment, and, therefore, banks continue to utilize the existing system. On the other hand, in the new field of information systems, risk management, and others, there are several examples of down-sized system installations. For example, distributed open systems are gradually increasing as replacements of mainframe computers. The reason for this increase is largely associated with short development terms and reduced cost.

1.1.3 Networks around the Banking Industry

The general structure of networks around the banking industry is shown in Fig. 1.3. The main network system is explained chronologically in the following discussion. The detail of the entire network system is described in the Center for Financial Industry Information Systems (2004a). The Domestic Funds Transfer System (Zengin system) began to operate in the age of the first-generation on-line system. This on-line network system was built up by financial institutions in an alliance with the Domestic Funds Transfer System. This is an on-line network system that deals with the transmission/receipt of messages concerning domestic exchange and the calculation of exchange settlements. Since April 1973, the starting time, the system has been reviewed and remodeled four times to meet with new requests.

The Society for Worldwide Interbank Financial Telecommunication system (SWIFT)[21] for the transaction of international banking was utilized from March 1981 in the age of the second-generation on-line system. It is an on-line network system that deals with messages regarding settlement and remittance of customers in international business of banks in alliance with SWIFT. However, SWIFT cannot settle like the Zengin system. The settlement itself is made between the corresponding banks and through the clearing system of each country.

The Credit and Finance Information System (CAFIS) was introduced in February 1984. The role of this system is to rationalize the transactions of credit card sales for member shops and banks. Members of CAFIS are able to access the CAFIS center and utilize all kinds of information services regarding sale transactions by using a Credit Authorization Terminal (CAT).

The Automatic Answer Network System for Electronic Request (ANSER) is the network that is utilized to manage transactions of electronic banking (EB). This is an on-line network system that deals with the request from the customer by audio response or some other means and connects banks with their customers.

The Bank of Japan Financial Network System (Nichigin Net) began to operate the current account system in October 1988. This is an on-line network

---

[21]SWIFT was established in May 1973. It is a nonprofit organization with headquarter in Belgium.

system that connects branches of the Bank of Japan with associated banking organizations. The objects of business utilizing the system are as follows:

1. Settlement of the current account, clearing, domestic exchange, foreign exchange in yen, and financial futures in yen.
2. The advance between the Bank of Japan and banking organizations.
3. The settlement about receipts and payments of bank notes and national debts.

Overall, the Nichigin Net is the very important for final settlements among banks.

The Credit Information Network, which began to operate in October 1988, has a database containing information gathered from banks, credit companies, and so on. This on-line network system is able to furnish the requests of members with the information contained.

The Funds Information Relay System (FIRST) began to operate in April 1989. The role of this system is to transact agreement messages for the collection and delivery from securities corporations and investment companies to trust banks. Its use promoted rationalization and sped up office work.

The Multi Integrated Cash Service (MICS) began to operate in February 1990. This is an on-line network system that connects with the MICS center in cooperation with the on-line CD/ATM system of each banking organization. MICS can relay the payment request of cash and the message of payment permission. It can sum up all debts and credits.

In addition, there are various other on-line network systems in the banking industry. These systems change and grow in relation to the diversification of banking business, and therefore, further reviews and modifications can be expected in the future.

1.2 Accidents That Affected On-Line Systems and the Lessons Learned

In this section, some notable examples of accidents that greatly impacted on on-line systems are discussed (Setagaya cable fire, 1992 London explosion, great Hanshin earthquake, Y2K problem), as well as countermeasures derived from risk management. The lessons that were learned from these incidents are also discussed and summarized. Chap. 2 discusses system problems that arose in 2002 following bank mergers.

1.2.1 Setagaya Cable Fire

Table 1.2 lists the main lessons that were learned from the Setagaya cable fire. This incident is one of the most serious accidents in the history of on-line banking systems. On November 16, 1984, an underground cable fire occurred at the Setagaya telephone office of the Nippon Telegraph and Telephone Public Corporation (now NTT), and all telecommunication lines were cut off within its control area. The importance of security policy was highlighted by the magnitude of the disruption to the on-line system.

Table 1.2. Lessons from Setagaya cable fire

Area of operation	Points of concern
Business administration	• Importance of contingency procedures • Education of staff regarding the fundamentals of bookkeeping • Careful administration of EB customers • Education of staff regarding on-line equipment
Business management	• Full explanation to customers about the circumstances of the accident • Redirection of customers to other banks according to their needs • Taking emergency measures (making notes for the future)
Information system	• Importance of the backup system • Readjustment of networks (duplicate lines, lines distributed about the route, wire and wireless networks, etc.) • Re-evaluation of the general terminal machine • The custody of backup data files
Others	• Securing communication methods (mobile computing and communication etc.) • Securing food, lodging, and goods for assistant members

At 11:50 a.m. on November 16, 1984, a Friday, the fire broke out from an underground passage in the Setagaya telephone office. The fire was accidentally started during an inspection by staff of a subcontracted company. As a result of the fire, 80000 telephone lines, including urgent lines to police and fire stations, were cut off within its control area. Customers were unable to use on-line terminals and CD/ATMs of banks in the area. Two nationwide banks that housed computer centers in the area were seriously affected by the accident. Office clerks of banks had generally become familiar with office work with the use of the on-line system. However, the accident suddenly introduced the need for office work to be completed by hand, and, as a result, there was a great deal of confusion at the branch windows after the fire. However, calm office processing could be performed gradually by the following actions:

1. Clerks and bank staff were able to obtain more accurate information about details of the accident and how customers who came to the branch office were affected.

2. They paid the money by paper documents of the balance account.
3. Staff took practical measures by checking the identification of customers.
4. Staff politely redirected customers to other banks according to their needs, particularly in the cases of those having accounts with other banks and money orders for other banks, and so on.

On November 17, the Saturday, the fire was put out 17 h after it began. Work then began in earnest to repair the damage. The on-line systems of banks were reconnected with the lines of branch offices and the systems were utilized gradually. Television and newspaper services informed the public about the accident and people were soon satisfied with the response of the banks, while correspondence with customers in the affected branch offices gradually calmed down.

In 1984, banks in Japan opened for Saturday morning business, and so the affected branches began their repair work after closing time. Severe discrepancies occurred between branch offices that were restored to normal conditions and others, through the incomplete classification of slips and checking of transactions.

In some cases, there was concern about the clearing system including the transaction of in- and out-clearings in each branch and the completion of comprehensive transfers and so on. Special mention must be made that general confusion did not occur owing to the overall support of the Federation of Bankers Associations of Japan (now the Japanese Bankers Association), the clearing house, and so on. However, customers of EB were inconvenienced during the period in which the lines were down, because they could not operate the money transfer service, the personal computer service, and telephone and facsimile services, etc. For account transfers made by concentrated transactions, it was necessary for banks that had previously obtained the consent of consignment receipt companies to draw on the basis of the balance of a central file and to delay money transfers slightly.

On November 18, the Sunday, the on-line systems of banks were significantly restored, and it was forecasted that CD/ATM services would be available in the following week. Fortunately, the day after the accident was a Sunday, and remedial office work was conducted in the branch offices to correct data entries that were not completed as a result of the accident.

Various lessons were learned from the Setagaya cable fire. Although the authors wrote that it was necessary for banks to reeducate employees on the foundations of bank bookkeeping, it was an opinion expressed after the event. Because such an accident in which the system stops for several days will rarely occur, banks considered that reeducation of staff was not justified.

The accounting systems of today are developed in accordance with the theory of bookkeeping. However, clerks of branch offices are largely engaged in operating terminal machines and do not take the principles of bookkeeping into consideration while on duty. They are not experienced in making the slip by hands under the on-line system and do not completely understand interoffice accounts, suspense receipts, and suspense payment, etc. Therefore, they are not familiar with the classification of account subjects. If the system is cut off for several days under such circumstances, heavy confusion will prevail in the branch offices. From this perspective, it is necessary to educate clerks with elemental bookkeeping skills and provide full training to branch office managers as a matter of routine. There is also a widely held opinion that a contingency procedure and manual are not effective, because accidents occur in ways that are beyond imagination. However, from a different viewpoint, one might argue that because accidents are not planned, it is necessary to learn from previous experience. It is recommended that banks actively utilize contingency procedures, not only when accidents occur, but also in the continual training of employees in emergency measures.

Although the security policies of the bank systems in this accident were different, much was done to address the issue of backup systems. The options available for backup measures included backup centers, backup machines, duplicate lines, lines distributed about the route, and the custody of backup files.[22] In addition, wireless network have also come into recent use. These lessons must be utilized for the system development and subsequent business administration.

1.2.2 The 1992 London Explosion

Table 1.3 describes the lessons that were learned from the London explosion of 1992. At about 9:20 p.m., April 10, 1992, a Friday, an explosion occurred in the international money market district in London. This incident, later identified as a terrorist attack, left three English citizens dead and many injured. Two Japanese banks that were located in the Commercial Union Building near the site of the explosion were seriously damaged. Urgent programs were made at branches and head offices directly after the incident and required the establishment of close communication with the head office in Japan. The banks took measures to address the safety and first-aid of staff, confirm the damage status, retrieve essential documents and goods, set up a temporary office, execute system backups, organize a support team, and establish communications with local authorities and those in Japan. Fortunately, the days following the blast, April 11 and 12, were a weekend and this allowed the measures described above to be implemented during that time. The Japanese banks continued business from their temporary offices on the Monday, April 13. The measures taken by the banks prompted the general view that all

---

[22]Regarding the duplication of communication lines, if duplicate lines were used to link the same bureau, this measure would provide no protection in an accident such as the one discussed here. Afterward, regulations were revised to allow multiple bureaus to be linked. It is desirable that essential lines such as those from Tokyo to Osaka are dispersed via different routes, e.g., via the Pacific coast district, central Japan, and the Sea of Japan coastal district.

Table 1.3. Lessons from the 1992 London explosion

Area of operation	Points of concern
Business administration	• Review of the contingency manual • Training in measures against disasters • Storage of essential documents in an outside custodial company
Personnel administration	• Maintenance of the staff safety and emergency measures • Organization and dispatch of assistant team
System	• Review of system backup (system, data, etc.) • Securing the backup office. • Securing communication devices (Internet, mobile terminal, etc.) • Adjustment of communication networks (formation of triangle and duplicate routes about networks)
Others	• Securing goods for staff and assistant members • Supply necessities of life • Connection to authorities and customers

had been done to maintain the order of the financial district. Their head offices in Japan acted as proxies for their London branches and made urgent arrangements for the computer downtime not to cause customers much inconvenience over the clearing of money and currency exchange. As a result, the banks were able to provide customers with necessary and minimum services until the Easter holidays. (April 17-20) The organizations were then able to effect complete restoration of services during this period.

With the progress of globalization, Japanese banks enlarged branch networks in various countries. However, in doing so, the banks also became exposed to new risks. These included industrial action of staff, possible lapses in law and order, terrorism and abduction, natural disasters (earthquakes, storms, floods, fires), environmental contamination, infectious disease, political disturbance, and war. At the occurrence of these types of events, there are many cases in which the bank might deal with the situation in connection with the jurisdiction department of head office.

There were many lessons from the London explosion such as how to maintain the safety of staff, how to maintain the backup system and communication, and how to provide essential services in collaboration with the outside companies.

1.2.3 Great Hanshin Earthquake

Table 1.4 lists the main items to arise from the great Hanshin earthquake. In the early morning of January 17, 1995, a Tuesday, the great Hanshin earthquake occurred, with the seismic center on the north area of Awaji island, close to Osaka and Kobe. Serious damage was inflicted on the Hanshin area. (metropolitan areas of Osaka and Kobe) About 6000 lives were lost in the disaster and 40000 houses were completely destroyed. The disaster also caused

Table 1.4. Lessons from the great Hanshin earthquake

Area of operation	Points of concern
Business administration	• Adjustment of contingency plans for disasters • Execution of training for measures against disasters • Arrangement of special measures and temporary steps (clearing, Zengin system and conducting business on holidays, etc.)
Personnel administration	• Confirmation of staff rules for emergency situations • Securing members and support • Securing transport and lodgings • Dispersion of the allocation between residences and companies
Information system	• Evaluation of the computer center and its ability to withstand an earthquake • Adjustment of network system • Utilization of the Internet • Securing devices for mobile communication
Others	• Installation of a private generator • Securing power supply car and transportation means • Making a variety of communication devices • Supply of the necessities for life • Utilization of alternative transportation routes

deaths and injury to bank staff in the affected area. Some branch offices in the affected area were unable to open for business because of the damage to public transport and utilities, damage to branch offices, and the outbreak of fires. However, the banking industry did its utmost to maintain the order of financial systems, and normal conditions were gradually restored without great confusion. This earthquake left various lessons to examine in the future. There were a number of problems in maintaining the safety of public transport and utilities. When the entire damaged area is paralyzed by the disaster, it is impossible for any one company to function effectively. Traffic facilities were cut off and could not be utilized for many days. Several routes and means were used to bring goods and essential items into the Kobe area. Means of transport included ship and motorcycle, while air transport provided air drops by passing over Kobe from Osaka to Okayama. Communication could be maintained by mobile telephone and the Internet, although disconnections occurred frequently as a result of the system networks being stretched beyond their capacities. Data transmissions of EB under this situation were effected by performing transactions via the computer centers in other areas.

In terms of the response of the banks, a number of special measures and temporary procedures were put into place after the disaster. For example, a number of clearing houses in the damaged area were temporarily closed and banks opened some branch offices for extended business hours. Any important or urgent business that was normally dealt with at branch offices that were closed by the disaster was handled at a central branch office. Banks also opened for business on holidays to provide services to those affected by the disaster.[23]

Written accounts have always proved to be invaluable sources of information for historical events, and Endo (1995) is considered particularly important in highlighting the impact of the great Hanshin earthquake on banking services. Endo was a branch manager at the Kobe branch of the Bank of Japan on the day of the disaster, and afterward wrote detailed accounts of his experience as a leader responsible for maintaining the order of the financial system. His book is precious as a real-life contingency manual.

1.2.4 The Year 2000 (Y2K) Problem

Table 1.5 lists various lessons that arose from the Y2K issue. The last few years of the 1990s saw considerable resources directed toward the Y2K problem. In the early development of computers and associated software, the year was described by two numerals that represented only the decade and the year number. The remaining two digits that represented the century were not used because the century was "understood." This system worked perfectly well during the 20th century, but as the year 2000 approached, fears were widely held that the two-digit code for the year 2000 would be misread as the year 1900. With the crucial importance of dates in the business data of banks, considerable measures were taken to address this anticipated problem. The year 2000 was an exceptional leap year[24] and there were problems of implanted chips that could not deal with normal conditions. Because of the widespread use of computing systems in society, the Y2K problem presented a challenge for almost all computer users. Governments of the world developed policies to promote a smooth transition and to instill a high level of public awareness. Subsequent cooperation between users, computer makers, and computer software developers eventually ensured that the feared social confusion was avoided.

Problems of various magnitudes that arose for banks dealing with the Y2K problem were as follows:

1. Banks incurred high costs in addressing the issue.
2. Banks stopped the development of new systems.
3. Demand for new products and software to replace old ones increased.

---

[23]See Daiwa Bank History Committee (1999) pp218-221.

[24]The year 2000 was a leap year, while the year 1900 was not.

Table 1.5. Countermeasures to the Y2K problem

Management behavior	• It was significant that the board did not leave the Y2K problem to only one organization but to the whole company. As a result, there was a thorough understanding of the information system.
Staff behavior	• The system department examined and checked the self-development program and the purchased software. • With the spread of end-user computing, user departments examined the software in use. It was then necessary to obtain the cooperation of manufacturers and software dealers. • In branch offices, they examined and checked hardware and software in use with the cooperation of electronic banking customers.
Risk management	• The system department and the user department each compiled a manual for risk management. • Plans were made for coping with the possible emergency risk.
Information system	• With the opportunity provided by the Y2K problem, software retained from the past was examined and checked. • Unnecessary computer software was disposed of.

4. Programmers familiar with COBOL, a language that was no longer widely used, were suddenly in demand.
5. Many staff were dispatched to offices to address oddities that arose in the change from 1999 to 2000.

In dealing with the Y2K problem, it was essential that company leadership did their utmost to stress the importance of the Y2K problem and to demonstrate the company policy toward it. System development relating to management strategy is becoming more and more important. This dictates a high level of involvement of company management in systems development.

1.3 Roles of the System Department

The banking system was changed greatly along with the improvement of online systems. Furthermore, system departments that were closely associated with this aspect of the banking business were also greatly changed. The following discussion explains the development process from an off-line system to that of today, and the challenges that face the system department.

1.3.1 History of the System Department

The punched card system (PCS) and computers used for batch-processing were installed in about 1960. The department to promote the mechanization of banking business was established at that time. Initially, this section was not an independent department and belonged to the department of general affairs within the bank. The section investigated the theme of mechanization, developed the systems required, and was responsible for the smooth installation and operation of the machines. At the time, many keypunchers were employed for data entry duties. Because of the number of staff, personnel management and health administration were important issues in the section.[25] However, with the commencement of the first-generation on-line system, banks gradually promoted the use of the on-line system for much of their business. The requirement for batch-processing reduced and keypunchers soon vanished from the office.

Three shifts of computer operators were required as the new service mode began, along with the operation of the on-line system. At the time, no other type of work in the bank required three shifts, although it is universal today. Persons in charge of computer operation were perhaps embarrassed in these early days because computer operation was regarded as blue-collar work. However, in time, the staff who did not take an interest in computer operation became burdens on their employers. It is conceivable that this matter was one of the reasons for outsourcing in the current department operation.

In the age of the first-generation on-line system, managers were appointed to business planning and business administration sections to supervise operation of the information system. Their main responsibilities were to rationalize office work and reduce the size of the labor force. At the time, the functions of the organization were not completely separate, and the person in charge of the information system was involved in all aspects, such as specification, programming, and testing. Programming language used an assembler and some programmers were artisan spirits owing to the delicate and specialized nature of the required coding. After on-line systems became established, their effects also became clear. Although the system department had a supporting role up to this time, it gradually became an essential function of bank operations.

In the age of the second-generation on-line system, the system departments of many banks were independent from the business administration departments. The system departments were gradually divided into the system planning section, system development section, and the system operation section, and so on. The number of staff in the system department increased drastically. Objects of the on-line system were rationalization or administration of the business as in the past, but the role of the system was also enlarged to the provision of detailed information relating to customers and business development. At the time, the development languages used were COBOL and

---

[25] The labor union raised concerns about protection against occupational diseases such as tenosynovitis and tendonitis.

PL/I of the compiler language except for the Assembler. The productive efficiency of system department gradually increased, although the costs of development still remained as a major expense. As the backlog of work for system departments continued to increase, decisions were taken to entrust systems development to outside companies. Initially, the systems development departments entrusted the subsidiary companies with only part of their operations. Afterward, further duties were assigned to outside companies. According to the enlargement of information system department, the operation department was also entrusted to outside companies.

In the age of the third-generation on-line system, the system development department fulfilled a role that addressed the needs of whole sections in the bank and was deeply associated with the core of the bank. The size of the systems development department increased and it became greatly influential within the bank. However, the needs of the system development began to exceed the available capability and the costs of the systems development department became excessive. With the development of the third-generation on-line system, a major theme for the systems development department was to entrust the outside companies with system integration. Furthermore, emphasis was placed on reducing the length of development periods that outside companies required. Outsourcing spread all over the world to be a commonplace form of systems development and reorganization of system departments advanced quickly. Various banks used outsourcing in regard to the system development and the system operation. As a result, although the system department became temporarily large, departments soon became somewhat leaner. With the development of the third-generation on-line system, outsourcing to other companies quickly became widely promoted. In doing so, two forms of outsourcing exist: one is the development of part of a system, while the second is the integration of an entire system. It is possible that the latter developed as strategic outsourcing. By the time the third-generation on-line system was running, most staff in the system department had entered the bank as general employees and had been assigned to the system department by normal personnel rotation. However, growth of skill in employees in the system department was strongly affected their suitability to the specific tasks of the system department. In general, staff needed to work much more independently than in the other main sections of the bank. Because of this situation, some employees worried about their futures and whether to choose generalist or specialist careers. However, more recent job seekers appear to have changed from the traditional company-oriented perspective to one that is occupation oriented. Therefore, banks now employ systems workers as required and regard them as different from staff in other banking sections. The training curriculum for the specialist, and the measures used for the treatment of problems have been adjusted to reflect this new attitude. After the third-generation on-line system came into operation, the system department became closely related with the development of management strategy. Chap. 2 will cover more on this topic.

1.3.2 The Changing Roles of the System Department

Generally, the business in charge of the system department is empowered to
address the following points:

1. Adjustment between the systems planning section and other related sections.
2. System development.
3. Establishment of the infrastructure for the information system.
4. The smooth operation and control of the information system.
5. Risk management for system security and crime prevention.
6. Training to promote utilization of the information system.
7. Evaluation of the information system.

The contents of information systems change greatly along with the progress
of information technology. Especially in modern banking, it has become common
for banks to be unable to promote management strategy without the
information system. In such cases, the ability of the system planning section
to make changes according to the management direction becomes more and
more important. To develop advanced and complicated systems, it is necessary
to obtain support not only from the computer maker and the computer
software provider but also from universities and research institutes. From this
viewpoint, it can be argued that the responsibility of supervising the system
and the role of the system department became increasingly important.

Important themes in system development have been to speed up the development
process and to reduce its cost. These points have been motivating
factors in outsourcing. Two banks introduced outsourcing in 1997 and another
12 banks followed suit at the end of 2000. Introductions of outsourcing were
spread in all directions, e.g., the development of computer software, cooperation
and combined utilization of the computer center including the backup
center, and utilization of the ATM network.

Given the combined efforts of a large number of staff in the system department,
and the efforts of computer makers and software providers, it is
remarkable that the improvement of documentation technology in the system
planning section became increasingly required. It became more and more important
to strengthen the system development power and to develop thorough
risk management strategies. The ability to manage the whole system was of
great importance with the prevalence of end-user computing and the spread
of open networks. The thoroughness of the information management never
failed to make an effect in every corner of the bank, although it is necessary
to research and grapple with security problems. Today, information management
must be accomplished by all parts of the bank, and not only by just
one section. For this reason, information systems are strongly supported by
auditing sections because of their ability to strengthen internal control.[26]

---

[26]See Hanaoka (1995) pp57-76.

Many smaller banks have not always followed the above-mentioned developmental steps, but have implemented a number of part measures. Various problems that emerged after the implementation of the third-generation online system for nationwide banks are discussed in Chap. 2.

Information System Strategy of Nationwide Banks

H. Nagaoka, Y. Ukai, and T. Takemura

2.1 Current Banks and Systems

The current banks may make a strong impression as if they devote themselves to various problems under severe financial circumstances, such as early write-offs of nonperforming loans and the speeding up of restructuring. However, it is clear that banks are eager to grapple with information strategies by using the newest technology for the development of business models conformed to the new age. The scope of these objectives extends to the entire management field. In such circumstances, it is imaginable that banks themselves will be faced with the age of innovation on a large scale.

In the 1980s, deregulation of the financial system gradually advanced in the banking industry. In the 1990s, securities, trusts, and insurance businesses gradually carried out mutual entry of other sectors of the financial industry and the period of strong administrative guidance by the government (the so-called convoyed system) came to an end. For example, the ministry letter that was used under the convoyed system by the former Ministry of Finance was abolished in July 1997. Financial products that did not win favor among customers became more popular, because troublesome office work that was required under the old legislation was no longer necessary. For example, debit cards entered circulation in January 1999 and drew public attention as a usable product.[1]

The development of new products and the management of new businesses were promoted actively along with changes of management conditions. In addition, management conditions of banks were greatly changed by the progress of information technology. In particular, it is said that the diversification of

---

[1]See Murata (1999) pp31-36.

delivery channels had a great impact on the management strategy of banks and contributed to the improvement and completion of customer services.

In relation to this, risk management concerning information systems is becoming a very important subject. In the management of operational risks, it is said that countermeasures for security and system problems are urgently required. In this chapter, the authors describe technical problems that need to be considered in the future and other problems related to system management.

2.2 Management Subjects of the Banking System

Table 2.1 shows the management subjects of information systems that concern the current banks. However, the division of main themes and items is not always clear, while some themes extend through a number of subjects. General views of eight management subjects are presented in the following sections.

2.2.1 Countermeasures for Changing Conditions

Along with the financial liberalization advances, the power of system development is required to cope with changing situations. The bank is closely dependent on the power of system development and whether it deals with a variety of business or makes some particular choices. With the entry of companies from other industrial sectors into the banking industry, it is expected that the competition between the banks and the newcomers will intensify, after which cooperation may occur.

2.2.2 Strategy for Competitive Advantage

In January 2001, the real-time gross settlement began to transact a settlement in the Bank of Japan. At the same time, it was also planned to strengthen the countermeasures of clearing risk. Afterward, it was expected that the hybridization of network settlement systems[2] would be promoted to.survey the examples of advanced countries.

In addition, the continuous linked settlement (CLS) is now being actively promoted. Through the special-purpose bank called the CLS bank, the CLS system is able to perform continuous clearing in real time for the selling and buying of currencies in foreign exchange transactions and thereby helps to reduce Herstatt risk.[3] A new system known as the Multi-Payment Network has

---

[2]Hybrid system implies the mixture of an on-time net clearing system and a real-time gross settlement system. See Shukuwa (2000) p53.

[3]When foreign exchanges occur, the settlement times in each country may be different if the transaction is made between time zones. Therefore, although a trader may send money in one currency, there is a possibility that funds may not be received in another currency in the destination country. This risk is known as "Herstatt risk." See Shukuwa (2000) p203.

Table 2.1. A list of main management problems, main themes, and items facing information systems

Management problems	Main themes and items
Environmental changes	Deregulation of monetary system (interest rate, branch office allocation, etc.), interaction between securities, trusts, and insurance companies, internationalization (global network, international cash management system, international cards, etc.), entry from other industries, and upgrading of clearing system.
Plans for competitive advantage	Development of high-technology monetary products, improvement of development power to new products, enlargement of delivery channels, enlargement of electronic banking, ATM strategy, and business model patent.
New technologies	Internet, cellular phone, electronic money, broadband, and several types of new media, etc.
Strategic support system	Strategic support to branches (one-to-one marketing), support system of head office (branch analysis, product analysis, profit management, etc.), cost management, and asset management.
Improvement of customer services	7 days/24-h services, consulting to asset management, in-store branches, ATMs in convenience stores, debit cards, IC cards, and digital certification.
Risk management	Assets and liabilities management, new BIS regulations, catching several types of risks (exchange, rate of interest, liquidity risk, etc.), safety measures, criminal measures, information security, and asset evaluation.
Thoroughness of business rationalization	Centralization of business transactions, low-cost operation, image processing (seal checking), digital filing, and multimedia. (multifunction terminal, mobile terminal, etc.)
Information system management	Functional change of system department, on- and off-the-job training of system development staff, priority of system development and investment, outsourcing, utilization of associated companies, cooperation with manufacturers and computer software providers, and system auditing.

emerged,[4] and it is planned to be promoted under the cooperation of many agencies. By this network, public charges and taxes can be paid through the use of various channels such as bank windows, ATMs, by telephone, personal computer (PC), mobile terminals, and so on. This system is more convenient for users than systems used previously and banks will be able to reduce burdens on business through its use. A particular merit of the system is that

---

[4]The Japan Multi-Payment Network Improvement Conference was held on May 11, 2000.

companies and local governments are instantly informed on transfer and receipt of funds. These settlement infrastructures are expected to be strongly promoted in the public welfare system in the future. Given the competition between banks, it is necessary for them to develop products that are attractive to customers. In addition, business strategies may also be changed by the diversification of delivery channels. Important strategic themes are the expansion of ATM networks and the promotion of electronic banking (EB) to regular customers. Furthermore, EB has the ability to serve individuals and households (home banking) and corporations and companies (firm banking). Table 2.2 lists the main services that are available at present. It is necessary for the product and business planning department and the information system department to cooperate closely to develop such fields.

2.2.3 Adjustment to New Technology

Planning the application of technological progress to business is a permanent and ongoing subject. The Internet banking and mobile banking are presently in the spotlight as new channels, while the age of broadband will see further development in this area. It is expected that new information terminals will become increasingly common. It is conceivable that electronic money will enter into practical use through electronic commerce (EC) after suitable testing is complete.

2.2.4 Strategic Support System

It is generally thought that the establishment of an information system for one-to-one marketing is a necessary development and that call center systems and telephone banking should be strongly involved. These two systems are an influential means to approach high-income customers. On the other hand, a main function in retail banking is to achieve mass sales. However, this type of sales is not beneficial unless it is operated at low cost. An information system to accurately calculate costs is essential for the development of a new channel and product. Moreover, the information support system and the network system for planning and administration are also essential.

2.2.5 Improvement of Customer Services

7 days/24-h banking have come to be normal aspects of ATM and nonbranch banking. However, the importance of providing a traditional channel for consultation over financial applications is also recognized. From this point of view, a "click and mortar" presence[5] remains essential for customer services.

---

[5]The expression of "brick and mortar" can be used to describe a company that conducts traditional business practices. The term is derived from the fact that the use of bricks and mortar is a conventional means of building construction. In contrast

Table 2.2. Main services of firm banking and home banking

Money collection service

Type of service	Function of the service
Improving efficiency of money collection	
Money transfer	This function automatically collects and distributes money between head office and branches. This service is useful for rationalization of money control to customers.
Automatic collection	This service directly draws school expenses, house rent, control charges, etc. from accounts of payers and transfers funds into the accounts of the receiver. This function reduces the labor force required to collect money.
Collection of bill payment	This is a contract which makes it possible to transfer money from the bill receiver's account to the bill sender's one.
Rationalization of collection business	
Administration of collection bills	This provides administrative materials about due dates, payers of bills receivable.
Advice of money received (this implies notice, communication, inquiry)	This service makes a list of details of money received on accounts. This is useful for customers to check the status of money received. This informs customers about money received on account and answers inquiries.

Money payment service

Type of service	Function of the service
Improving efficiency and strictness of payment office work	
Payment control of public fees	Automatic withdrawal of public fees can be checked. It is essential for companies that have many branches and offices to manage public fees.
Drawing up a letter of transfer request	This service can draw down a fixed amount of money from an appointed account regularly and pay the receiver. The labor force required for office work can be reduced.
Automatic transfer (approval by magnetic tape, facsimile)	This can draw down a designated amount of money from a nominated account regularly and pay the receiver. It is designed to reduce office workforce.
Alternative of making bills payable	This service can reduce accounting section employee in the requesting company.
Improving efficiency of accounting and the general affairs department	
Payment of local taxes	This substitutes the payment of local taxes for members of a customer's company.
Payroll calculation	This substitutes the calculation of payroll for customers and pays salary into the designated bank.
Cashless system of expenditure in companies	This can pay costs (travel allowance, etc.) into designated accounts.
Offsetting accounts receivable and cash accounts	This service can send advice of money received to the customer and offset accounts receivable and cash account automatically.

Along with the new types of branch offices that have appeared in recent years, the general concept of the branch office has also changed. Various types of branch offices, such as in-store branches (i.e., branch office facilities in convenience stores), in-branch stores (shops in the bank branch office), and joint or cooperative branches (branches of a bank and a trust bank or similar in a single office), have appeared with the progress of restructuring. Furthermore, card services have also changed with the availability of debit card services on a cash card. This advance is noteworthy, especially when considering the likely impact of the introduction of IC cards.

2.2.6 Risk Management

Risks that concern the banking industry are credit risk, market risk, liquidity risk, business risk, and system risk. However, our discussion is restricted to business and system risk, or so-called operational risk. The execution of a new BIS regulation was initially postponed for 1 year from 2004. According to the Nihon Keizai Shimbun on April 26, 2004, the execution of the new BIS regulation was forecasted to be postponed for a further 1 year from 2006. However, it is necessary to establish the operational risk whatever may happen. Operational risk naturally includes system security and crime prevention; however, it is essential to organize a stricter administrative system than the present one for dealing with risk in the open network. Staff who deal with sensitive information tend to be spread into all parts of the bank by gradual dispersion. Therefore, it is important for the bank to be comprehensive in its information administration policy. The study of policy against cyber-terrorism was advanced by the Information Security Policy Promotion Congress established in January 2001. Communication and operation systems were organized together by government and private business. On the other hand, the police organized a security policy conference in October 2001 to counteract the menace of cyber-terrorism, and is also establishing the organization for emergency communications and the restraint of damage in close cooperation with important infrastructure enterprises.

In this movement, the systems of the banking industry are also important as social infrastructure. The various risks associated with the computer system are shown in Fig. 2.1. It is necessary to delicately control the countermeasures used for safety considerations and crime prevention.

2.2.7 Maximizing Business Efficiency

With the continual changes in technology and the demands of customers, it is impossible for business reformers to reach a point where no further reforms can

---

to this, the expression of "click and mortar" is now used to describe banking firms that gather customers by both the Internet transactions and through traditional branches. See Nakano (2000) p187, and Nishigaki (2000) p159.

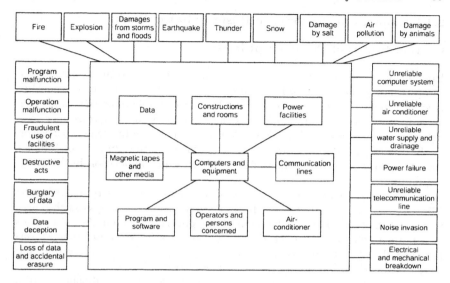

Mitsui Marine Safety Service Department (1993) p4 with modifications

Fig. 2.1. Sources of risk surrounding the computer system

be made. Therefore, it is necessary to embrace the advances of new information technologies and terminal machines and strive to maximize efficiency. In doing so, much of the business that was traditionally conducted in the back office in the branch is absorbed in concentrated transactions by the business center. After this, branch offices are not only places of office work, but are also ones for sales and business consultation.

2.2.8 System Management

A continual challenge for bank staff is to meet the diverse developmental needs of the bank in a timely fashion with the use of limited management resources. However, it is unrealistic for complete systems to be developed or revised by a single bank and at its own expense. It is preferable to reduce costs and quicken the system development by outsourcing. It is also important for banks to educate staff about the improvement in the bank systems and to retain staff with relevant specialist skills.

2.3 Revolution of Delivery Channels

There are many management subjects that relate to the information system. Let us consider the diversification of delivery channels that has reformed the point of business in the bank.

2.3.1 Changes to the Point of Business in the Bank

Current Status of the Point of Business

Previously, banks obtained permission from the former Ministry of Finance
to open new branch offices. For example, by the business notice of the Ministry of Finance, there were several kinds of branch offices at that time such
as ordinary branches, mini branches, mechanized branches, consumer loan
branches, robot-retailing branches, agencies, and mobile branches.[6] However,
in the 1990s, the administration of branch offices by the Ministry of Finance
was changed in order to promote independence and competition in the banking industry.

In 1995, the regulation of established branch offices was retained by the
Ministry of Finance, but the establishment conditions for new branch offices
were abolished. In addition, the regulations concerning the number of persons
in a branch office were abolished and the classification of mini branches was
removed. Robot-retailing branches were finished by the report system and
were expanded from the view of the business strategy.

Although the process of deregulation progressed, the establishment of a
branch office still required permission in accordance with the Bank Act Article
8.[7] However, on November 2, 2001, the Revised Act of part of the Bank Act[8]
was finalized, and the establishment of the business office was changed to
require only the giving of notice.[9]

The current banks have promoted restructuring and rationalization such
that ordinary branch offices have become much leaner operations. Based on
different functions and business strategies, different types of branch offices
have emerged.[10] Examples are as follows:

---

[6]See Editing Committee for Handbook of Bank Regulations (1990) p238.

[7]Japan Bank Act Article 8. (Establishment of a Business Office): If a bank intends to establish a branch or other business office, to change its location (including
a change in the main office), or to alter its type or abolish it, the bank shall, except
as otherwise designated by an Ordinance of the Cabinet Office, obtain the approval
of the Prime Minister in a manner prescribed by an Ordinance of the Cabinet Office.
The same shall apply when a bank intends to establish or abolish an agency.
Refer to the Japanese Bankers Association (2001) Appendix IV, p25.

[8]Refer to Bank Act Enforcement Regulations Article 10. This regulation has
laid down an example that does not require approval about the establishment and
abolishment of branch office or change of position.

[9]Refer to the Japan Bank Act Article 8. (Reform Bill): The contents are as
follows. Today, a bank has to submit notice (change/installation of a branch office,
and a position) to the Prime Minister of Japan. Note that the bank had to obtain
the approval of the Prime Minister when it would intend to establish a branch, to
change its type, or to abolish it in a foreign country, formerly.

[10]See Sugimura (1999) pp86-118.

1. In-store branch: Banks may take note of large stores such as supermarkets that have many customers and then seek to establish a branch office in the store.

2. In-branch store: Banks may allow retail stores or coffee shops to operate in the branch office with the aim of benefiting both parties.

3. Kiosk branch or mini branch: Small-sized branch offices that provide limited services to customers.

4. Multimedia or robot-retailing branch: Branch offices that provide services by ATM, automatic consultant machine, or automatic contract machine (ACM), PC, and so on.

5. Mobile branch: Banks put ATMs and another machines on their car or bus in rural area.

6. Total financial services branch (one-stop banking services branch): These branches are able to provide total financial services to customers in the same branch office, e.g., opening deposits, credit card business, purchase of goods, and so on.

7. Joint or cooperative branch: A branch office of one bank is established within a branch office of another bank in a cooperative arrangement, e.g., a bank and a trust bank may operate within the same facility.

8. Network of hub-and-spoke type branch offices: A group of branch offices made up of unmanned branch offices and special type ones, with a central mother branch acting as a core in the area. (see Fig. 2.2)

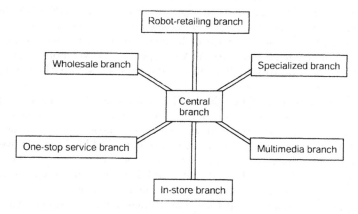

Sugimura (1999) p117, and the Center for Financial Industry Information Systems (2000a) p287 with modifications

Fig. 2.2. Hub-and-spoke type branch networks

In classifying the character of the branch, 1 and 2 describe the relationship between the owner of the branch office and the tenant; branch types 3-6 describe the special features of the branch; and 7 and 8 describe the nature of branch integration in the overall banking service.

The background of improving the diversification of branches is affected by the supporting information systems such as ATMs, ACMs, multimedia terminals, and so on.

The general trend of office work in current branches has been affected by the absorption of simple business by ATMs and network banking. The scope of activities in the back offices of branches has been reducing by centralization to the business center and EB. As a result of this trend, branch offices are becoming increasingly minimal and specialized.

Development of the ATM

The development of the ATM network is important in the strategy of banking business. Currently, banks are building their ATM networks by installing off-premises machines in accordance with their own judgment, although it can be seen that many ATM installations are examples of cooperative arrangements. For instance, banks in the same area mutually utilize ATMs in cooperation with other banks. Banks utilize ATMs in 24000 post offices all over Japan in cooperation with Japan Post. Furthermore, banks installed ATMs in convenience stores in cooperation with their management companies (see Table 2.3), and did the same in many railway stations.

In particular, it is thought that convenience stores are favored by ATM users because of ATM availability at night, their location in safe accessible areas, the provision of parking, and so on. One kind of ATM used in convenience stores uses 2000-yen notes to reduce the space required in the store and to increase the money held. Ordinary ATMs dispense 1000-yen notes and 10000-yen notes. In relation to this, it is interesting to observe that the circulation of 2000-yen notes is increasing.

In the installation of off-premises ATMs, traditional banks have struggled to secure suitable sites with adequate safety provisions at reasonable cost. However, according to the plans of the main convenience stores, as shown in Table 2.3, banks will rapidly increase the number of ATMs in stores over the next 2 or 3 years. Such installations are already secure and present favorable conditions for use of the machine. Robot-retailing branch offices in convenience stores provide a convenient service to customers, and, while branch offices of banks have mechanical guard systems in place, there is no security staff present. It is considered that the presence of staff in the convenience store is desirable for the control of safety. The ATM network will make rapid progress through the cooperation of convenience stores and banks. As this occurs, the original promotion plan of each bank will be forced to be changed.

Table 2.3. ATMs in main convenience stores

Convenience store	Management company	Equipment plan	Number of establishments	Main cooperative banks
Seven-Eleven	IY Bank	3650 stores by spring 2002, and 7150 stores by 2005[a]	About 2420 in January 2002, and 9213 in August 2004[b]	UFJ, Yokohama, Sumitomo Mitsui, Resona (Asahi until 2003), Shinsei, Tokyo-Mitsubishi
Lawson	LANS	3000 stores by spring 2002, and 4000 stores by 2003	About 1700 in January 2002, and 3312 in July 2004[c]	Tokyo-Mitsubishi, Sumitomo Mitsui, Resona (Daiwa until 2003), UFJ
Family Mart	e net	5000 stores by spring 2002	About 3640 in January 2002, and 5107 in July 2004[d]	Tokyo-Mitsubishi, Sumitomo Mitsui, Sumitomo Trust and Banking, Mitsubishi Trust and Banking
am/pm Japan	@ bank	1274 stores under the improving plan	About 1140 in January 2002	Sumitomo Mitsui, Japan Net

a) Includes Ito Yokado Co., Ltd
b) http://www.iy-bank.co.jp/atm/atmnet.html (accessed August 3, 2004)
c) http://www.lawson-atm.com/atm.area/ (accessed July 31, 2004)
d) http://gis.e-map.co.jp/standard/12130010/index.htm (accessed July 31, 2004)
Miyao (2001) p85 with modifications

2.3.2 Network Banking Strategy

New delivery channels for banking services have gradually emerged to complement traditional branches. The use of these new channels is explained in this section.

Telephone Banking and Telemarketing

Telephone banking has become possible through the advancements made in telephone and computer technology, and provides various services such as account inquiry, remittance, money transfer, opening accounts and canceling of contracts, consultation, and so on.

According to the authors' interview with a call center of former Daiwa Bank in 2000, there are various reasons why customers use telephone banking. Some of these reasons are as follows:

1. It is not necessary for customers to go to a branch office, allowing them to save much time.

2. Customers are accustomed to using telephones. Some are more at ease with the telephone than with a personal computer because they can verify transactions by their own voice.
3. Telephone banking is convenient for transactions that require specific timing according to currency exchange rates.
4. Telephone banking can be used after the closing time of business, although this depends on the type of service required.
5. Customers can use telephone banking when they cannot reach a branch office, when they are working, or when away on business.

There are two types of responses available to customers using telephone banking. One is the interaction with a computer-controlled audio response, while the other is direct communication with a telephone banking specialist. For the latter of the two options, several points need consideration as follows.

1. Organization of the necessary personnel
2. Training of staff to become familiar with the range of services offered by telephone banking
3. Security and confidentiality of customer information

In contrast, telemarketing is a business activity of the call center for the sale of strategic and promotional products to customers. It deals with the promotion of new goods, sales to customers, fixing money from liquidity deposits, reminders of due dates, and so on. For effective telemarketing, it is essential that a database of customer information is established along with a system that can easily accept inquiries regarding sales information.

Internet Banking

A research institute in the United States has released survey results concerning the operational costs in banking. This survey clearly showed that the cost of providing office staff to deal with customers at the window is the highest, and that the cost of providing services by the Internet is the lowest.[11]

It is likely that the situation in Japan is the same. The Internet banking does not require branch facilities and associated staff. Costs can then be held low, although a number of indirect costs need to be considered. Banks are strongly promoting the utilization of the Internet banking in the retail field as a low-cost delivery channel. However, some misgivings in making a fast transition to the Internet banking have been expressed as follows:

1. Middle-aged and older citizens tend to have more wealth and property, but may have little or no experience in using the Internet. Therefore, it is not good business management to promote a systematic strategy that excludes wealthy customers.

---

[11]See Department of Commerce (1998) p29.

2. Most users of the Internet are young men. Therefore, there may be considerable time before such customers mature and gain significant wealth. Until such a time, the Internet banking services are unlikely to return a profit.
3. Internet banking is a convenience rather than a necessity, because there are no banking products that change from hour to hour.
4. Is the security of the Internet transactions adequate?

In considering the first two of these concerns, it is likely that the number of customers using the Internet banking will gradually increase, such that the service will begin to return profits. However, it is probably necessary for banks to promote the Internet banking as a convenience for customers, as part of its diversification of delivery channels. Banks that provide the Internet banking services will have a marketing advantage over those that do not. This is important at a time when it appears that the number of users of the Internet banking is gradually increasing.

The use of 7 days/24-h banking has now become widespread and it is used by many people who find it inconvenient to use conventional banking services. (dual-income households, shop proprietors, single parents, office workers, etc.) It is frequently used on holidays and at night through the Internet because of the convenience to the customer. Banks offer various new services to the users of the Internet banking, although it is apparent that the "cash-based society" of Japan remains strong and will persist for some time yet. Thus it is reasonable to expect that considerable time will be required before the use of the Internet banking attains the usage levels of ATMs and credit cards. Furthermore, although more information leaks occur via traditional banking services, some people remain concerned about the security of personal information when using the open infrastructure of the Internet. Therefore, an ongoing information campaign is required to enlighten and reassure customers. Although banks may provide special favors to users to help expand the use of the Internet banking, it is also necessary to determine whether customer desire to use it is increasing.

There are two specialty services that banks provide. One is the Internet banking, and the other is virtual banking. One aims at getting new customers, while the other aims at increasing the use of the Internet and virtual banking by regular customers. The investment amount of the latter is supposed to be less than the former.[12] These two movements are notable after this. Anyway, in the Internet banking, it is an important subject in future to develop new strategies utilized the bidirectional specialty.

Mobile Banking

Mobile banking is utilized for account inquiries, transfer deposits, and foreign currency deposits by using a cellular phone with a web browser. It is a new

---

[12]See Osaki and Iimura (2001) pp205-218.

delivery channel, and the business and services available are frequently re-newed. Cellular phone terminals that can be utilized include i-mode, EZ-web, and J-Sky (now Vodafone!) among others. However, a point of contention that has already risen relates to the incompatibility of the communication protocols used by each phone.[13] Therefore, it will be necessary for banks to meet the requirements for all kinds of terminal machines. Currently, the third-generation cellular phone service (IMT-2000) is already in operation, and, at the moment, it is very difficult to forecast the extent of the use of this service.

If the use of this new service becomes widespread, with the ability to deal with data such as moving pictures, it should see more practical use as the next type of multimedia terminal. The system uses the same infrastructure that is used in foreign countries after the adoption of an international standard. This should lead to advances in the quality of the communication, which, in turn, should see the acceptance of new banking services as technical improvement occurs.

2.4 Creation of Mega Banks and System Problems

In 2002, two serious incidents related to banking systems occurred in the Japanese banking industry, and resulted in near paralysis of banking services for large numbers of customers. Both situations arose from mergers of financial groups and were precipitated when the integrated systems were brought into operation.

On January 15, 2002, the service of automatic debt transfer in UFJ Bank, a subsidiary of UFJ Holdings, inc.. was badly affected on the first day of the merger; about 1.75 million withdrawal transfers from UFJ Bank accounts were postponed and about 180000 withdrawals were doubled. Prior to this disaster, the initially planned time frame for the merger was shortened to just 3 months. This meant that the time available for integration of the system was also shortened 3 months. The process was a landmark event in terms of banking systems: there can be no greater trial than the operation of a newly integrated system on the opening day of a new, merged bank. The integration process was required to deal with such complicated business as the reforma-tion and synthesis of the computer center, and the synthesis of the ledger file and so on. Regrettably, and despite much effort, serious system troubles occurred. Afterward, the adequacy of the integration period was considered to be questionable.

The second major disruption of financial services occurred on April 1, 2002, when Mizuho Bank, a subsidiary of Mizuho Holdings, inc., was estab-lished. System problems occurred from the opening day, and these included the partial operation of ATMs, denial of cash card payments, inability to with-draw funds (about 2.5million incidents), so-called double withdrawals of funds

---

[13]See the Center for Financial Industry Information Systems (2000a) p192.

(about 60000 accidents), delay of money transfer, doubling of remittances, and delays in the information about money received.[14] This accident occurred on a wide geographical front in Japan and affected a large part of society. In the wake of Mizuho meltdown, the Financial Services Agency (FSA) issued a business improvement order based on the Japan Bank Act, Article 26.[15] It required measures for improvement to banking systems, countermeasures to prevent similar reoccurrences, and clarification of the responsibilities held by banks. It is shown in detail on the home page of the FSA (http://www.fsa.go.jp/). Furthermore, the FSA identified the causes of this accident as insufficient system tests and operational tests, poor communication within the development organization, and an incomplete business infrastructure. It was also concluded that the company board was lacking in understanding the risks involved in the merger process and this led to critical delays in decision making.

Regarding computing systems, it became not uncommon for banks to nominate a preferred computer manufacturer. Three mega banks did this shortly after mergers. The bank of Tokyo-Mitsubishi decided to adopt products by IBM, Sumitomo Mitsui Bank adopted products by NEC, and UFJ Bank[16] adopted products by Hitachi. In opposition to these three mega banks, Mizuho Holdings, inc. was the focus of attention in the banking world over whether a particular computer manufacturer would be selected. Computers and their applications used for accounting systems of the three merging banks were different from each other; then Fuji Bank had used products by IBM, then Dai-Ichi Kangyo Bank had used products by Fujitsu, and then Industrial Bank of Japan had used products by Hitachi. Naturally, it is assumed that much effort was required to effect the merger in the given circumstances, but despite

---

[14]See Nikkei Computer (2002) pp8-74.

[15]Refer to Japan Bank Act Article 26 (Suspension of Banking Business) : The Prime Minister may, when deemed necessary to maintain the sound and appropriate management of banking business of a bank in light of the business and/or financial conditions of the bank or the financial conditions of the bank and subsidiaries, etc., require the bank to submit an improvement plan detailing the measures that it will take and the deadlines for doing so in order to ensure sound management of the banking business, or order the bank to modify an improvement plan already submitted, or to the extent it is necessary, order the bank to suspend its banking business in whole or in part until a predetermined date, or order the bank to transfer its assets to the competent authorities, or take other such measures as needed for supervisory reasons.

When the orders pursuant to the provisions of the preceding Paragraph (including orders to submit improvement plans) are deemed necessary in light of the capital adequacy of the bank or of the bank and its subsidiaries, etc., said orders shall be the orders specified by an Ordinance of the Cabinet Office and the Ministry of Finance based on the categories of capital adequacy of banks or of banks and their subsidiaries, etc., determined by an Ordinance of the Cabinet Office and the Ministry of Finance.

Refer to the Japanese Bankers Association (2001) Appendix IV, p34.

[16]UFJ Bank will be merged into another mega bank by the end of Oct. 2005.

these efforts, a lack of understanding and leadership in the top management was the fundamental cause of the ensuing difficulties.

Various questions arose in the wake of the Mizuho disaster:

1. Did a department that was responsible for the unification of systems exist, and if so, did it have adequate resources and the backing of company management such as that seen in dealing with the Y2K issue?
2. Was there sufficient communication among each group in the project?
3. Did the company board fully consider the opinions of the system department?
4. Did the company consider the risk of opening business on April 1, with the added volume of data associated with the end of a month and the end of a fiscal year?

Despite the problems of the Mizuho disaster, the experience led to a new understanding of the importance of systems management and served as a warning to other banks that such trouble could also entangle their operations if the same mistakes were made. It is recommended that all banks strengthen their departments responsible for risk analysis. It is likely that it will become increasingly necessary to employ the use of outside consultants and teams for system auditing. Lessons from the troubles described above were reported by Kamiyama (2002), Nose (2002), and Miyamura (2002).

2.5 Issues

The improvement of information technology is continuous. As a result, the state of information technology is deeply connected with the information strategy of the bank. Important technical and management issues related to the progress of information technology are discussed in Subsects. 2.5.1 and 2.5.2.

2.5.1 Technical Issues

Introduction of IC Card

The IC card was invented in France in 1974 and is not new technology. Its notable advantages are the ease of introducing high levels of security, the card itself is capable of recording considerable volumes of data, physical damage to the card is minimal because no contact is required for operation, and it can be used as a medium of electronic money. However, IC cards have been scarcely utilized in Japanese banks in comparison with European and American banks. A main reason for this is that Japanese banks had already issued about 300 million cash cards by the end of 1999 and use of the cards has taken root in society.[17]

---

[17] See the Center for Financial Industry Information Systems (2000a) p34.

In replacing existing cash cards with IC cards, banks could not help but hesitate due to the high cost of issuing IC cards incurred by measures such as the reconfiguration of terminal machines, the maintenance of computer software, and the necessary staff time involved in the changeover. On the other hand, the credit industry has continued to test the use of IC cards. A particular advantage of the IC card is its high level of security and this has become an important issue with recent increased use of forged cards.[18]

In the banking industry, it has become standard practice to issue a single card that can act as both a cash card and credit card for the sake of customer convenience. Not surprisingly, the concept of using an IC card as a multifunction card has received attention. Reasons for the increased interest in IC cards are as follows:

1. The need for a personal identification number has increased with the spread of debit card usage and the increase in fraudulent card use.
2. In testing of the use of electronic money, the use of IC cards has proven successful.
3. The use of IC cards has been adopted in various industries.
4. The necessity of having to use different cards for different financial services has prompted interest in identifying a more convenient system.

One bank has actually issued multifunction IC cards and others have begun to issue cash cards as IC cards. The use of debit cards was initiated in January 1999 and was considered a significant move toward a cashless society. The debit card system has various advantages for the consumer, such as:

1. There is no need to carry cash.
2. If consumers already hold a cash card, there is no need to make a special application for the debit card facility.
3. There is no need to pay a fee when the debit card is used.
4. The level of spending is limited by the amount of the deposit.

However, it is essential that the password is used with the utmost care when using the debit card. Although cases of fraudulent use are rare, consumers are protected by insurance from losses incurred by burglary and forgery, and by establishing a spending limit. The use of IC cards as a solution for such problems is a subject worthy of investigation.[19]

It should be noted that the standard specifications for cash cards and their development are different from those for credit cards. The former are governed by the Japanese Industrial Standards (JIS II type) and the latter are governed by the International Organization for Standardization (ISO; JIS I type).[20] To promote the use of IC cards as multifunction cards, it was considered necessary

---

[18]See UFJ Institute (2000) pp112-122, and Iwamura (1996) p26.

[19]See Research Group of Financial IT (2000) pp153-156.

[20]See Segawa (1993) p27, and the Center for Financial Industry Information Systems (2000a) p9.

to integrate the operating systems in various industries and to establish a uniform standard for the infrastructure of the card industry. The Japanese Bankers Association established the IC Cash Card Standard Specifications in March 2001. In the developed countries of Europe and in the USA, IC cards have already come into widespread use, and it is considered that financial transactions in Japan are likely to be rapidly improved by such widespread use of IC cards.

The establishment of standardized specifications provides an impetus to change from the conventional cash card to the IC card. Such movement would minimize wasteful operating costs, reduce the system development costs at the time of conversion to the IC card and new terminal machines, and improve the level of convenience to the customer. Because the system is based on an international standard, it is expected to have a favorable impact on future business, and such effects are expected to continue along with technical improvement.[21]

Access to Broadband

Upgrading to an access line called "last one mile" is essential for broadband. Until the installation known as "fiber to the home" (FTTH)[22] is achieved, cable television service (CATV), ADSL, wireless, and satellite facilities are utilized to access broadband services. At present, the competition between providers of ADSL services to sign new customers has become very intense, particularly in the home market.

CATV was introduced ahead of asymmetric digital subscriber line (ADSL) despite its higher charges. While ADSL connects through a conventional telephone line, CATV requires the installation of specialized cabling and therefore incurs significantly higher setup costs. In addition, CATV has been restricted to business areas. These factors have combined such that the number of ADSL customers has increased by more than the number of CATV customers. In September 2001, ADSL service providers began serious price competition and savings to customers increased as the competition intensified. By the summer of 2002, the charge for a regular ADSL service became the lowest in the world. Furthermore, the introduction of cheap wireless broadband communication services has seen renewed intense competition among the alternative broadband infrastructures, which should continue on until the age of FTTH.

With the availability of broadband services, terminal machines can send and receive considerable data at high speed. It is therefore likely that the variety of products and the diversification of functions will be dramatically promoted in network banking. It is notable that wireless phone technology

---

[21]See Otsubo (2001) pp18-23.

[22]By completely converting the access networks of users to optical fiber, consumers can send and receive various types of information such as voice, image, and animation, data at high speed on broadband lines. See Nakano (2000) p44.

such as IMT-2000, the next generation mobile communication system, and digital television technology are used in terminal machines. Because of the decision to move to terrestrial digital broadcasting, the conversion from analog television to digital appears likely, and it is conceivable that the demand will amount to about 17 trillion yen.[23]

However, the extent of uptake will determine whether the unit price is cheap or expensive. It is considered that digital television will be an important information terminal in the broadband age after its establishment. If broadband service is utilized by the Internet, wireless telephone, and digital television, home banking services will become increasingly popular. If the restrictions of time and location can be reduced by the diversification of delivery channels and an increase in the level of functions, the age of "ubiquitous banking[24]," in which financial services can be provided at all times and everywhere, will soon occur.

2.5.2 System Management

Reaction to Operational Risk

Banks become exposed to certain operational risks with the spread of the Internet banking. Prior to this, in the age of the mainframe computer and leased line network, system engineers could deal with system problems according to fixed procedures. However, when trouble occurs in the open network system, it is difficult to assign responsibility to any particular section or entity, such as the bank, the communication company, the provider, the system vender, and so on. If the bank requires much time to restore the computer system, its reputation in the public domain is damaged.

New regulations of BIS propose that banks should come to terms with operational risk posed by accidental loss and fraudulent business practices. Upon the installment of new technology, the bank must consider measures to fully control the risk related to system capability, data archiving, guarantees of security, and so on.[25]

In addition to obvious risks, there are various issues that may be difficult to deal with in actual system management. Examples of these more sinister risks include file deception, confidentiality breaches, computer viruses, unlawful invasion, wiretapping, denial of service (DoS), and so on.[26] Banks will normally establish an information security committee to protect against these

---

[23]If analog broadcasting can be completely converted to digital broadcasting in the 10 years from 2000, as expected, the scale of market for receiving terminals will reach a total of 164.897 billion yen over 10 years and each antenna should be responsible for business that is worth about 1.281 billion yen. See Nishi (2000) p51.

[24]Ubiquitous implies being in existence everywhere. See Miyazaki and Kitamuro (2001) pp81-84.

[25]See the Center for Financial Industry Information Systems (2000a) pp528-532.

[26]See Yamaguchi (2001) pp81-127.

attacks. However, although many offenders do not have highly technical means at their disposal, crimes by fraudulent use of information continue to occur frequently. After copies were made of the seals on customers' bankbooks by using a scanner, crimes occurred at branch offices in Japanese banks with the use of the forged seals. In response, banks have developed a system to register and administer the seal as electronic data and then planned to abolish the copy of the seal in the bankbook. Although this system incurs costs to develop and implement, it contributes not only to protecting against fraud, but also in rationalizing office work at the window. Mischievous and fraudulent crimes are a kind of shadow against the light of the information society. Plans need to be in place to deal quickly with such problems when they arise.

Enhancement of Potential Development

There are many business activities that concern the system department of a bank, such as the opening of new business, the development of new products, strengthening of system management, the adoption of new technology, and so on. It is natural to invest in system requirements according to their priorities, but it is also necessary to keep the development terms of any new measures to a minimum. Development time is closely connected with cost, and forms an essential part of any business strategy. The predominant position in a competitive business also depends on how quickly good systems are developed.

Outsourcing of the systems section is a remarkable means of strengthening development power. In the early stage, outsourcing was enforced in the operational department. In the development department, it was normal to entrust system development to the computer software provider and the computer manufacturer. Recently, outsourcing of the system department has become a strategic policy, because it reforms the organization itself, strengthens the developing power, and aims at reducing the cost. Three possible scenarios of outsourcing the system department are as follows:

1. The bank separates the system department and reestablishes it under a separate budgeting system as an independent profit center. The bank entrusts this subsidiary company with the system development, which, in turn, also entrusts system development to an outside company. In this case, all or a majority of the system department personnel may be transferred to the subsidiary company. An important theme in this practice is the maintenance of high technical skill levels in the subsidiary company.

2. The bank leaves the system planning section within its organization, and enlists the computer manufacturer and the computer software provider to conduct the development business. The systems development section in the bank is scaled down and its members are transferred to other sections. It is anxious that know how of the development and the technical capability disappear from the inside of own company to entrust outside company with the system development. It is necessary to fully understand the risk of outsourcing.

3. The bank, the computer manufacturer, and the computer software provider
   make a joint corporation by mutual investment. The bank entrusts the
   joint corporation with the development business, which is conducted by
   personnel from each company. The bank leaves members of the systems
   planning section and transfers members of development section to the joint
   corporation. It can combine the business knowledge of the bank with the
   technical power of the computer manufacturer and the computer software
   provider, and can speed up system development. However, it is necessary
   to manage the personnel of the combined team to ensure that the levels
   of communication and trust between team members remain high.

In any case, it appears that there is a limit to which the bank should
strive toward in strengthening the system development. It is essential to make
a joint company in some form with outside companies such as the computer
manufacturer and the computer software provider, and to be closely connected
with each other.

It is extremely important that suitable personnel be trained and educated
to deal with system management. As such, training and the education are very
important, although companies cannot adequately follow progressive technical
innovation by employee education alone. Because it is difficult to recruit skilled
engineers from other companies in Japan, it is necessary to request the skills
of suitably qualified personnel from outside companies. This tactic is regarded
as an essential condition of strategic outsourcing from the view of the system
department. A clear line can be drawn between simple outsourcing, which
merely entrusts outside companies or subsidiaries, and strategic outsourcing in
which a strategic alliance is formed. In addition to this, there are recent cases
in which various banks have worked together to develop computer software and
system development was entrusted to an outside company. In the bank-related
business, there are examples of banks working together to utilize a computer
center, including a backup center, and leasing the facility, while entrusting
an outside company to operate the center. Furthermore, several banks have
cooperated to utilize a joint ATM network. In such circumstances, a case can
be made to entrust an outside company with ATM administrative business.
Through such outsourcing, banks can strengthen the development potential of
information systems and rationalize business and administrative operations.

Adaptation to Business Model Patent

It is noteworthy that banks are now taking measures to defend their intellec-
tual property (IP) that has emerged in the process of developing new financial products and business models. Until now, the banks had sought patents
for original products when the information systems were developed in-house.
There was a case in which a bank and a computer manufacturer, which dealt
with system development, made a joint patent application to cover the development of the computer software and computer hardware. The merit in

obtaining the patent lies in the ability to then sell the technology to another bank. This represents an enlargement of the market for the computer manufacturer and enables the bank recover part of the development cost. However, such cases are rare.

A business model patent was granted after a patent dispute between Signature Financial Group Inc. and State Street Bank and Trust Co. In 1998, Signature Financial Group Inc. was granted a patent for its general purpose computer system for the investment trust, but State Street Bank and Trust Co. claimed that the patent was invalid. In July 1998, the Court of Appeals for the Federal Circuit (CAFC) rejected this claim, and the business model patent was admitted. In December 1999, the inspection rule similar to that used in the USA came into force in Japan.[27]

Naturally, if a bank can obtain a business model patent faster than its competitors, it stands to gain an advantage over other banks. Therefore, it is important for banks to protect company IP and obtain patents before others. In this age, banks channel many resources toward a wide range of project areas, and, as a result, hold a great number of computer software assets. It is an important aspect of management strategy to raise the value of these assets by patent acquisition. However, it is doubtful that the group responsible for the technology in question should merely instruct a separate department to proceed with a patent submission. The bank should provide an effective management strategy for patent acquisition that sees the technical group, the business planning section, and the patent attorneys work together closely.[28] Examples of relevant success stories in the USA include the "One-Click Method" by Amazon.com., and "Inverse Auction" by priceline.com. However, it is to be regretted that a business model that occupied a market share in Japanese industry has not been recognized. Therefore, banks must establish a long-term view and provide systematic organization from an early stage to reduce delay in this field. This is an important and essential subject in management strategy. In Chap. 9, the contribution of computer software assets to the company value is discussed and estimated.

System Auditing and Corporate Governance

In 2002, serious system problems occurred successively in UFJ Bank and Mizuho Bank, a subsidiary of Mizuho Holding, inc. The avoidance of any repetition of these incidents is now an important management subject. Given the huge scale of integration and business development that occurs in the merger of a mega bank, it is not surprising that system accidents should occur. However, a key issue is how to limit the effects of such problems so that they cause only minimal damage. The serious risks that need to be considered are shown as follows:

---

[27]See Koda (2000) pp42-52.
[28]See Y Matsumoto (2001) pp93-100.

1. Problems originate in the information system. Such problems may arise during the development of a new information system, during a large-scale system modification, after the renewal of system machines and terminal machines, after renovation of the network, or as a result of relocation of the computer center.
2. Problems arise as a result of changes within the organization. Changes known to have caused disruptions include rapid enlargement of the business, the release of new products, opening of new business, establishing or closing of branch offices, personnel changes, and a move to outsourcing. Included with such changes should be the reorganization of the total risk management system.
3. Issues arising from staff conduct. An example of this is the information leak that occurred in the system department of the Ground Self-Defense Force in August 2002. Such affairs related to personnel are increasing along with the progress of information technology. There are two problem types that can be classified as personnel risk: one is intentional crime, the other the occurrence of careless mistakes. It is essential to maintain consistent administration of personnel and the rigid discipline in the company to protect against crime. In the latter case, common examples of carelessness may be the loss of laptop computers, the copying of confidential documents, careless control of important papers, and so on. In addition, it is necessary to carefully manage the liquidity of trained staff as employees are being retired through restructuring. It is becoming necessary to filter business administration practices and to monitor private use of company assets, and so on. In addition, an important part of ensuring acceptable staff conduct lies in a strong emphasis on training in the company. As an aid to finance and banking companies, the Center for Financial Industry Information Systems issues the annual report of computer system troubles and crimes in financial institutions as a special issue of "Financial Information System" every November.

Risk associated with the computer system can exert a heavy effect on banking management and plans must always be at hand to improve the corporate governance. In addition to its primary function, an information system audit also serves to strengthen the control of the internal organization.

In July 1999, the Financial Supervisory Agency (now the Financial Services Agency) composed a manual used to inspect financial instructions. This required the risk management strategy from the banking company to be made available after the inspection by the authority. Thus, the banking industry was required to make an effort to establish sound and proper business practices. Every financial institution used the Center for Financial Industry Information Systems (2000b) as the auditing guide.

In the above discussion, the role of information system audit also was pressed to manage along with the current movement. The following five points are necessary for the system audit to function effectively:

1. An auditor who can issue exact instructions.
2. Communication with the executive board concerning reform. The board must take an active role in the system improvement.
3. The auditor must have strong technical skills and an ability to keep pace with technical progress.
4. The auditor must plan to utilize outside auditors for issues that cannot be addressed from within the bank.
5. Passive auditing will occur in times of normal operation, but it should become more demanding when important management subjects are reviewed or planned to ensure that active opinions should emerge.[29]

---

[29]See Kimura (2001) pp181-216.

Part II

Review of Information System Analyses

Limit of Aggregate Level Analysis of Information System Investment

S. Watanabe, and Y. Ukai

3.1 Introduction

In January 2001, the Headquarters for Promotion of Advanced Information and Communications Society (IT Strategy Headquarters) announced the e-Japan aim of Japan's becoming one of the most advanced nations in information technology within 5 years. (the 2001-2005 period)

In March 2001, the e-Japan plan which materialized the e-Japan strategy was announced. In this plan, the following five measures were listed to facilitate the formation of the advanced information and communications network society:

1. Construction of the world's most advanced information and telecommunications network
2. Promotion of education and learning, and the development of human resources
3. Promotion of e-commerce, and similar high-tech services
4. Computerization of public administration and promotion of the utilization of information communication technology in the public field
5. Safety and credibility in advanced information and communications networks

In the meantime, strengthening of the network had been quickly tackled in the United States since 1998, toward the end of the Clinton administration. The USA, France, Germany, UK, and EU all announced action programs on IT (information technology) in 2000. This movement was because the sustainable economic growth of the USA in the1990s was widely considered to be due to advances in IT.

Many studies have been carried out on whether IT is a leading cause of rapid rises of productivity and relaxation in the business cycle. Studies of

IT policy have also been performed in Japan by government agencies and academic institutions. Economic and business studies of IT in Japan and the USA can be classified into the following five main groups[1]:

1. Demand-creation effects
2. Positive productivity, cost cutting, and firm value effects
3. Employment effects
4. Consumer-surplus effects
5. Organization reform effects and management strategy effects

IT arouses a demand for IT-related goods with a demand-creation effect, and it spreads to a demand for other goods and boosts growth.[2] However, there is no guarantee that firms increase a demand for IT-related goods, if firms do not have the number of advantages of using IT.

Therefore, it is necessary to quantitatively analyze the effects of positive productivity and cost cutting, and increases in firm value. In the USA, the estimate of production and cost functions with IT are widely carried out.

In addition, there seems to be two employment effects of IT: a decrease in employment as IT capital replaces labor, and an increase in employment in existing and new industries by a demand-creation effect. It is possible to say that IT creates jobs at an aggregate level, if employment reduction in individual firms is skillfully absorbed by employment extension of other industries.

Moreover, not only the supply side but also the consumer-surplus effect are able to lower the price of IT-related goods. In addition, analysis of the business organization effect by IT has been carried out in recent years. We also review the management strategy effects on IT investment.

Part II presents a review of studies on the IT effects on firms such as productivity rises, cost cutting, and rises in firm value. In the 1980s, the relation between IT and productivity became a target of academic discussion. However, the effects of the astonishing advances of computer hardware and computer software on productivity were difficult to evaluate, leading to the so-called productivity paradox. The productivity paradox is an expression used to describe the situation in which IT might not contribute to productivity. However, in the 1990s, new data and the qualitative adjustment of data began to reexamine the relationship between IT and the productivity. By this improvement, studies in which IT has positive effects on productivity would be reported in great numbers. In addition, the approach diversified from analysis of the relationship between IT and the productivity to the contribution of IT to the market value of firms.

---

[1] A note on the e-Japan strategy and action plan is shown on the home page of the Prime Minister's Office: http://www.kantei.go.jp/jp/it/network/.

[2] In the classification by Workshop on Information and Communication Policy (2002), the wide-range effect like the change in consumption behavior has been taken up rather than the consumer-surplus effect.

In this chapter, economy-level and industry-level research on productivity and IT are examined, and a discussion of preceding research on the productivity paradox in the USA is presented. The same analysis is performed at firm-level in Chap. 4. Because the technical terms used by researchers of IT-related investment are different, in Part II, we unify the name into IT investment. The authors describe the content of IT investment, if necessary.

3.2 Productivity Paradox

The productivity paradox of IT is the phenomenon that IT capital does not lead to an increase in output. A famous phrase of Solow (1987) summarized it: "You can see the computer age everywhere but in the productivity statistics." The studies of productivity paradox grew steadily in the USA in the 1990s.[3] In initial studies, there was insufficient evidence that IT had a positive effect on productivity in the 1970s and 1980s.

Oliner and Sichel (1994) maintained that because the depreciation period of the computer is short in contrast with other capital, and the ratio of IT capital stock to total capital stock is much smaller than the ratio of general capital stock to total capital stock, the contribution of IT capital to economic growth becomes difficult to quantify.[4] This scenario is described by the small share of computers in capital stock hypothesis.

Steiner (1995) indicated that IT promotes international specialization of labor and the productivity effect has leaked out into foreign countries. In addition, it appears that due to inadequacies in statistical procedure, IT stock itself was not accurately reflected in statistical data until the latter half of the 1990s. The mismeasurement hypothesis of the data was carries out by Siegel (1994, 1997), and Siegel and Griliches (1992).

In addition, there is the long learning lags hypothesis that the IT effect to the productivity does not appear in the productivity when the accumulation of human capital does not advance, because the players that use IT capital stock efficiently are employer and employee. For example, the work of Brynjolfsson and Hitt (2000) examined in Chap. 4 shows the importance of time-lag. Besides, the offsetting factors hypothesis that an IT effect has not appeared at the aggregate level, is also considered in spite of the absolute IT effect at the firm-level, as shown by Berndt and Malone (1995).[5]

---

[3]Shinozaki (1998) estimated IT investment of Japan on the demand side from the input-output analysis. Kuriyama (2002) analyzed the production inducement effect on the final demand in IT industries.

[4]There is already criticism that the growth of input cannot explain the growth of output since the 1950s. See Solow (1957) for details.

[5]Kumasaka and Minetaki (2001) insisted that the interpretation of the cause of the productivity paradox by Oliner and Sichel (1994) will be wrong, if the productivity that Solow showed is the total factor productivity. The contribution of IT capital to economic growth and labor productivity is affected by the distribution of

There are firm-level studies that IT investment brings about considerable productivity increase in the USA. These studies are reviewed in Chap. 4. Oliner and Sichel (2000) verified that computers (computer hardware, computer software, and peripheral devices) greatly contributed to economic growth in the 1996-1999 period in comparison with the 1991-1995 period in their industry-level analysis. They also indicated that the contribution of computers to growth in labor productivity from the 1991-1995 period to the 1996-1999 period accounted for about 66% of the total growth. In this analysis, computer software data made available to the public by the National Income and Product Accounts (NIPAs) of the Bureau of Economic Analysis (BEA) were used.

Thus, the productivity paradox has not been verified with improving data and estimation methods in the latter half of the 1990s.[6] In this chapter and in Chap. 4, research on the economic effects of IT is arranged between the USA and Japan.

3.3 Studies of the Whole Economy

Table 3.1 shows the main studies of the economic effect of IT investment for

Table 3.1. Studies of IT investment effect on the whole economy

Method	Studies
Growth accounting	Oliner and Sichel (1994, 2000), Jorgenson and Stiroh (1995, 1999), Matsudaira (1997)[J], Y Ito (2001)[J], Ministry of Internal Affairs and Communications (2001)[J], Jorgenson and Motohashi (2003)[J]
Production function	Shinozaki (1998)[J], Shinjo and Zhang (1999)[J], Shinjo (2000)[J]
Cost function	Lau and Tokutsu (1992), Shinjo and Zhang (2003)[J]
Consumer surplus	Bresnahan(1986), Hitt and Brynjolfsson (1996)

[J] Studies on Japan

IT capital. However, because the total factor productivity is the value obtained by subtraction of the contribution of production factors from the economic growth rate, if economic growth is constant and the capital contribution lowers, the increase in the total factor productivity will become large. See Kumasaka and Minetaki (2001) pp43-44.

[6]Sichel (1997) classified the causes of productivity paradox into six main groups. This classification does not contain the international specialization hypothesis by Steiner (1995). The classification in Sichel (1997) includes the mismanagement hypothesis and the redistribution hypothesis except for four kinds of classification shown in the text. See Sichel (1997) for detail.

the whole economy. When the productivity effect of IT investment is measured, the methods for estimating the contribution to economic growth using growth accounting and estimating the productivity using a production function are taken.

3.3.1 Growth Accounting

Oliner and Sichel (1994, 2000) and Jorgenson and Stiroh (1995, 1999, 2000a) analyzed the contribution of the production factor to economic growth and labor productivity using growth accounting.

A standard analytical method using growth accounting is explained here. Three neoclassical assumptions - production function with constant return to scale, existence of competitive equilibrium (marginal product $=$ user cost), and non-existence of external economy effect - are used. Output Y at some particular time t will be a function of the stock of IT capital $,K_c$, the stock of other capitals K_o, its labor force L, product price P, rental price of IT capital r_c, rental price of other capital r_o, and wage rate w.

The Cobb-Douglas production function is:

$$Y = F(K_c, K_o, L, t). \tag{3.1}$$

Total differentiation of Eq. 3.1 with respect to time and substitution of the competitive equilibrium conditions $\partial F/\partial K_i = r_i/P$ for $i = c, o$, and $\partial F/\partial L = w/P$ into the equation give Eq. 3.2.

$$\frac{dY/dt}{Y} = \frac{r_c K_c}{PY} \frac{dK_c/dt}{K_c} + \frac{r_o K_o}{PY} \frac{dK_o/dt}{K_o} + \frac{wl}{PY} \frac{dL/dt}{L} + \frac{dF/dt}{F} \tag{3.2}$$

It is possible to decompose economic growth rate into the products of the share $(r_i K_i/PY)$ and the factor growth rate $((dK_i/dt)/K_i)$ in each production factor (contribution) and the growth rate of total factor productivity. $((dF/dt)/F)$

It is possible to calculate total factor productivity using the economic growth rate, each factor share, and each factor growth rate, because this equation is identical. However, it is necessary to obtain the user cost of computers in order to obtain the share of IT capital.[7] The share affected by the setting method of user cost would change the contribution of IT capital drastically.

Oliner and Sichel (2000) substituted labor force L with qL with quality q in Eq. 3.1. Jorgenson and Stiroh (2000a) used IT service in place of IT stock. This means that IT capital K_c of Eq. 3.1 is replaced with the sum of the service of IT stock considering aged deterioration.

---

[7]Wilson (1995), Brynjolfsson and Yang (1996), Nakaizumi (1998), and Brynjolfsson and Hitt (2000) are useful survey studies. Oliner and Sichel (2000) compared their estimates of the contribution of computers to economic growth in the USA, in the latter half of the 1990s, with those of other studies.

The effect of IT capital on economic growth rate is required from the contribution of IT capital (the product of IT capital share and IT capital growth rate) and the contribution ratio of IT capital. (a ratio of the contribution to economic growth rate) The effect of IT capital to economic growth rate is also evaluated by adding the growth rate of total factor productivity to the contribution of IT capital. Total factor productivity can be divided into technical progress in the computer supply industry and technical progress in the computer application industry.

We can rearrange the expression by subtracting the growth rate of labor from both sides of Eq. 3.1 to observe the contribution of each production factor to labor productivity.

$$
\frac{dY/dt}{Y} - \frac{dL/dt}{L} = \frac{r_c K_c}{PY} \left(\frac{dK_c/dt}{K_c} - \frac{dL/dt}{L} \right)
$$
$$
+ \frac{r_o K_o}{PY} \left(\frac{dK_o/dt}{K_o} - \frac{dL/dt}{L} \right) + \frac{dF/dt}{F} \tag{3.3}
$$

The history of the analysis of Japan-USA on growth accounting is explained here.

Oliner and Sichel (1994) estimated that the contribution of IT capital to the growth rate in gross domestic product (GDP) was 0.31% and the contribution ratio of IT capital to GDP growth rate was about 11.2% in the USA for the 1970-1992 period.[8] In their analysis, the contribution ratio and the contribution of IT capital were 0.25% and 7.31% in the 1970-1979 period, and 0.35% and 15.42% in the 1980-1992 period, respectively. It was concluded from this result that both the contribution and the contribution ratio of IT capital to economic growth were small in the 1980s.

In addition, Jorgenson and Stiroh (1995), in their analysis of the USA in the 1947-1992 period, reached the following conclusion: the contribution of IT capital to economic growth rate lowered from 0.52% in the 1979-1985 period to 0.38% in the 1973-1992 period, and the average growth rate of total factor productivity also lowered to 0.47% in the 1973-1992 period from 0.78% in the 1979-1985 period.

However, the contribution (the contribution ratio in the parentheses) of IT capital to economic growth rate was 0.49% (16%) in the 1974-1990 period, 0.57% (21%) in the 1991-1995 period, and 1.10% (23%) in the 1996-1999 period in Oliner and Sichel (2000). The contribution to labor productivity was 0.44% (32%) in the 1974-1990 period, 0.51% (33%) in the 1991-1995 period and 0.96% (37%) in the 1996-1999 period. In addition to this, the estimated total factor productivity was 1.2% in the 1996-1999 period. The estimated period and labor quality were different among studies with the positive effect

---

[8]The user cost is often set as $r_c = [r + \delta_c - \pi_c] p_c T_c$, where r is the net profitability of the capital, δ_c is the depreciation rate of computer, π_c is the capital gain rate, p_c is the computer price, and T_c is the adjustment term of tax.

of IT capital and the computer software data differed from that of Oliner and Sichel (1994).

Matsudaira (1997) found that the contribution of IT capital of Japan to real GDP growth rate was 0.38% and the contribution ratio of that was 11.3% in the 1974-1993 period. Although the contribution of the IT investment to real GDP growth rate in the 1974-1983 period was about 9.3%, it increased to 13.2% in the 1984-1993 period. This result is similar to that of Oliner and Sichel (1994). When the contribution of IT capital of Japan to economic growth rate and the contribution ratio exceeded the value of the USA (before 1983), IT capital was relatively more important in Japan than in the USA. However, after 1984 IT capital was relatively more important in the USA than in Japan. Matsudaira thought IT was better utilized for office automation and factory automation in Japan than in the USA in the1970s.

In the meantime, the contribution of IT capital of Jorgenson and Stiroh (2000a) was smaller than that reported by Oliner and Sichel (2000), because the definition of output was wider than the one of Oliner and Sichel (2000).[9] However, the contribution (the contribution ratio in the parentheses) of IT capital was 0.399% (14.5%) in the 1990-1995 period and 0.755% (16%) in the 1995-1998 period, being in agreement with the result of Oliner and Sichel (2000). Total factor productivity was 0.358% in the 1990-1995 period and 0.987% in the 1995-1998 period. That is to say, there would be a positive view on the economic effect of IT capital, when the analysis is carried out for the latter half of the 1990s.

Gordon (2000) indicated that the labor productivity in the 1995-1999 period varied from 2.8% to 0.6% when the factor of the business cycle was removed. He indicated that the effect of IT capital on labor productivity was apparent because the most of the increasing labor productivity in manufacturing industries other than the computer manufacturing industry was based on the extension phase of business cycle. However, Gordon (2002) insisted that technical progress and the recovery of productivity by IT capital after 1995 brought about economic growth.

Y Ito (2001) analyzed the contribution of IT capital on the Japanese economy. The point which greatly differs from the study of Jorgenson and Stiroh (2000a) is that the contribution of IT capital to economic growth rate maintains almost 0.2% in the 1990-1995 period and the 1995-1999 period, and 0.5% in the 1980-1985 period in Japan. The contribution of IT to total factor productivity was 1.8% in the 1995-1999 period. The contribution of IT cap-

---

[9]Oliner and Sichel (1994) used three kinds of definitions of IT capital. To begin with, the first definition is computer and peripheral equipment. Typewriters and business machines were added to the first definition in the second definitions of IT capital. The third definition of IT capital added communication equipment and copying machines to the second definition. The contribution shown in text body was calculated according to the third definition. They analyzed the contribution of computer service to economic growth rate using software, hardware, and the computer-related labor as the computer service.

ital to labor productivity hardly increased in the 1990-1995 period and the 1995-1999 period.

The low labor productivity was because the positive contribution of total factor productivity supplemented the negative contribution of lowering general capital stock after the collapse of the bubble economy.

The Ministry of Internal Affairs and Communications (2001) estimated the Cobb-Douglas production function with IT capital, and obtained the contribution and the contribution ratio of IT capital to economic growth rate. The contribution (contribution ratio in the parentheses) of IT capital to economic growth rate was 2.24% (45.6%) in 1985-1990, 0.64% (45.7%) in 1990-1995, and 1.23% (100.8%) in 1995-1999. The contribution ratio of IT capital in this research is considerably larger than that of Y Ito (2001).

Jorgenson and Motohashi (2003) adjusted Japanese IT data according to the USA definition to compare the IT contribution to economic growth between Japan and the USA in the latter half of 1990s. They found that the contribution of IT capital service to economic growth in the USA was similar to that in Japan in the 1995-2000 period. (USA: 1.1%, and Japan: 1.08%)The low rate of economic growth in Japan was due to negative contribution of labor service in the same period. (USA: 1.35%, and Japan: −0.2%)

The method of the factorial analysis of economic growth rate is greatly dependent on three neoclassical assumptions shown in Eq 3.1. In the analysis by growth accounting, the adjustment cost of investment does not exist.

The contribution of the capital may be estimated to be small when the share of the capital in which the adjustment cost exists is calculated only using the user cost, because the adjustment cost drives up the investment cost. Therefore, it is impossible to gain economic effect of IT investment accurately using such a simple macro method.

3.3.2 Macro Production Function and Cost Function

Lau and Tokutsu (1992) estimated the trans-log cost function with IT capital in the USA in the 1961-1990 period. As the result, half of the real economic growth has depended on the growth of computer capital.

Shinozaki (1996) estimated the Cobb-Douglas production function with IT capital in the USA in the 1977-1994 period .Although the marginal productivity of general capital stock was 0.2%, that of IT capital was 63.9%.

Shinozaki (1998) also analyzed IT investment in Japan for the 1974-1996 period using the same technique. Shinozaki concluded that the marginal product of general capital stock was at the gross rate of 25.2% and that of IT capital stock was at the gross rate of 136.0%. The marginal product of general capital stock was at the net rate of 17.0% and that of IT capital stock was at the net rate of 120.2%. Using general capital equipment ratio (general investment/general labor) and IT equipment ratio (IT investment/general investment) as explanatory variables, Shinozaki attempted the factorial analysis of labor productivity. As the result, it was confirmed that IT equipment did not contribute very much to the increase in labor productivity in the 1976-1994 period.

Shinjo (2000) estimated the production function with the gross domestic product for the USA and Japan. The data used in the analysis of IT capital of the USA were derived from Shinozaki (1996), and that for Japan was estimated in Shinjo and Zhang (1999) for the 1970-1995 period. The Japan-USA comparison examined the contribution of IT capital to economic growth rate. In this analysis, the gross capital stock was used as capital input. It was because they considered that the productivity of capital is faithfully shown by the gross capital. From the estimated result, the contribution ratio of IT capital stock in Japan greatly fluctuated in range of 20%-25%, and it was proven to be very much more unstable than the contribution ratio of IT capital stock in the USA. (about 30%) The marginal rate of return of IT capital stock in Japan is always bigger than that of IT capital in the USA. However, the marginal rate of return of both countries has declined over time. The marginal rate of return of IT capital stock in Japan is always larger than that in the USA. This first result is consistent with Shinozaki (1996).

Shinjo and Zhang (2003) constructed the log-run data of IT capital stock for the 1974-1998 period to compare IT effect of the USA to that of Japan. In addition , Shinjo and Zhang estimated the marginal benefits-costs ratio (marginal Tobin's q) of additional IT investment from the estimates of the short-run dynamic factor demand model for the 1976-1997 period. The result showed that $1 of IT investment reduced more than $1 of variable costs for the 1990s in Japan in contrast with the result of the USA. Shinjo and Zhang interpreted that the under investment in IT capital came out in Japan for the collapse of the bubble economy in the 1986-1989 period. Shinjo and Zhang also indicated that IT capital and non-IT capital were substitutes for the estimated period in Japan and the USA.

3.3.3 Effect on Consumer-Surplus

The studies of the consumer-surplus exist as a third index for measuring the IT effect.

Bresnahan (1986) analyzed the price decline effects of mainframe computers in the financial services sector (banking, brokerage, insurance, and related business)in the USA in the 1958-1972 period. It was estimated that the consumer-surplus became five or more times the expenditures at the 1972 price.

Hitt and Brynjolfsson (1996) estimated the consumer-surplus increase for the whole economy by studying the IT stock price decline. Hitt and Brynjolfsson concluded that IT stock price decline produced a consumer-surplus of 360 hundred million dollars per year and 1449 hundred million dollars in total in the 1988-1992 period. They also showed that IT did not bring out the firm profit, although IT raises the consumer-surplus and productivity of firms.

Unfortunately, a study that analyzes the IT effect on the consumer-surplus in Japan has not been conducted.

3.4 Industry-Level Studies

Table 3.2 shows industry-level studies in the literature on the economic effects of IT investment. It is necessary to decompose the whole economic growth into the industry-level in order to evaluate IT effects more precisely, because negative productivity growth in one industry may offset positive productivity growth in another. We review typical industry studies by dividing overall industries into manufacturing industries including IT production and non-manufacturing industries utilizing IT.

Table 3.2. Industry-level studies on economic effect of IT investment

Method	Studies
	Overall industry
Growth accounting	Jorgenson and Stiroh (2000a, 2000b), Y Ito (2001)[J]
Cost function	Minetaki (2001)[J]
Production function	Lee (1998)[J], Shinjo (2000)[J], Shinozaki (2001b)[J], Kuriyama (2002)[J]
	Manufacturing industry
Growth accounting	Siegel and Griliches (1992), Siegel (1994, 1997)
Cost function	Morrison and Berndt (1991), Morrison (1997)
Production function	Berndt and Morrison (1991, 1995), Berndt, Morrison and Rosenblum (1992)
	Non-manufacturing industry
Others	Roach (1991)

[J] Studies on Japan

Roach (1991) compared the productivity of non-manufacturing industries (a proxy for the productivity of service industries) with that of manufacturing industries in the USA for the 1950-1989 period. His analysis showed that service industries with low labor productivities used more than 85% of the installed base of IT. The low labor productivity in service industries can be mainly explained by over employment of white-collar employee. Roach insisted that IT capital increased white-collar employment in service industries, although IT capital replaced labor in manufacturing industries. These results showed that the productivity paradox existed in the USA.

There are studies that have confirmed the productivity paradox in manufacturing industries. Morrison and Berndt (1991) estimated the dynamic cost function by considering price and quality of IT capital, and concluded that computers do not have a strong cost cutting effect. In addition, Berndt and Morrison (1991) analyzed the production function model using the data of BEA including all USA manufacturing industries. He estimated the benefit-cost ratio of IT capital and other capitals. The result showed that marginal productivity of IT capital was positive. They confirmed overinvestment of IT capital because the benefit-cost ratio of IT capital was under 1 in most industries and smaller than the ratio of other capitals.

Berndt, Morrison and Rosenblum (1992) estimated the production function using the data of 20 manufacturing industries in the USA in the 1968-1986 period. The result showed that the correlation of computer and total factor productivity was weak in most industries in the 1980s.

Berndt and Morrison (1995) examined the relationship between IT (high-tech office and information technology) capital and total factor productivity in manufacturing industries in the 1968-1986 period.[10] The result showed the negative correlation between total factor productivity and increases in the share of IT capital in the total physical capital stock. Industry-level studies of the USA do not deny the existence of the productivity paradox.

There is the problem of measurement error in the above research. Siegel and Griliches (1992) analyzed measurement error in TFP. By improving the industry-level data, Siegel and Griliches showed a positive correlation between total factor productivity and average ratio of computer expend to capital expend in the USA during the period 1973-1986.

Siegel (1994) indicated that the producer price index (PPI) of the Bureau of Labor Statistics (BLS) underestimated, on average, 40% of quality change. In this index, it is assumed that the improvement of the quality is carried out only in the case in which the cost change is accompanied or for cases in which new goods are incorporated. In short, the quality change by the technological change, which does not accompany the price change, is not considered. Siegel showed that IT investment correlated positively with both labor quality and product quality.

Siegel (1997) formulated a model that controlled measurement error of quality to solve the problem in which the price index cannot take account of

---

[10]Hiromatsu, Kurita, Kobayashi, et al. (2000) analyzed industry-wide growth accounting. They showed the contribution of information equipment in the 1990-1994 period declined markedly in comparison with that in the 1986-1990 period. They concluded that the contribution of information equipment in the 1990-1994 period did not contribute to value-added growth. In the industry-level analysis by the Economic Planning Agency Coordination Bureau (2000), the six industries of transport machinery, electrical machinery, precision instruments, construction industry, wholesale trade, and retailing were shown as industries in which computer stock contributed to total factor productivity. Hiromatsu, Kurita, Tsubone, et al. (1998) analyzed the contribution of information equipment stock to labor productivity and the production function and showed that the contribution and efficiency of information equipment in the first half of the 1990s were smaller than those in the1980s.

quality improvement. He used industry-level data of the four-digit classification in the 1972-1987 period to show that the benefits of computers spread among industries. The result showed the positive correlation between productivity growth and IT investment. Therefore, measurement error played major role in the productivity paradox in the manufacturing industries.

Morrison (1997) estimated the cost function with the adjustment cost using industry-level data in the 1972-1991 period. He calculated Tobin's q of IT capital and non-IT capital from the estimates. The estimated result showed the overinvestment of IT.[11]

Lee (1998) estimated the elasticity of production with respect to IT capital in the cross section using the data of IT capital stock of 23 Japanese industries including the non-manufacturing industry. The data of IT capital were calculated from the input-output table for 1975, 1980, 1985, and 1990 by the technique like Shinozaki (1996). With his analysis, it was confirmed that the marginal rate of profitability of the IT capital was also considerably larger than the marginal rate of profitability of general capital over the entire period in Japan.

Shinjo (2000) estimated the Cobb-Douglas production function with IT capital and non-IT capital in 21 manufacturing industries of the USA using cross-sectional data from 1977 and 1992 to compare with the research of Lee (1998) on Japan. Shinjo showed that marginal profitability of IT capital was larger than that of other capital, although there was a problem of multicollinearity between explanatory variables.

Industry-level studies of the USA changed from a skeptical attitude toward the economic effect of IT investment to a standpoint that conceded an inability to confirm the economic effect of IT because of measurement error.

Jorgenson and Stiroh (2000a, 2000b) estimated the growth accounting equations for 37 industries of the USA in the 1958-1996 period. Jorgenson and Stiroh calculated the industry-level contribution to aggregate total factor productivity growth using the Domar weights (the ratios of industry output to aggregate value-added). The results showed that the contributions to TFP were positive in industries such as Electronic and Electric Equipment industry, while the contribution were negative in industries such as Services, Construction, Government enterprise, and so on. Two third of USA industries contributed to the aggregate TFP in the estimated period.

Y Ito (2001) analyzed how the industry-level productivity growth affected total factor productivity for 22 Japanese industries in the 1980-1998 period using the technique like Jorgenson and Stiroh (2000a, 2000b). Y Ito showed that IT capital contributed to the increase in total factor productivity in

---

[11] Within the BEA data, the office, computing and accounting machinery (OCAM) capital was used. Communication equipment, scientific and engineering instruments, photocopiers, and related equipment were added to the OCAM capital for analysis. These approaches are consistent with the definitions of the second and the third definitions of IT capital of Oliner and Sichel (1994).

industries such as electrical machinery, apparatus, commerce, finance, and chemistry. Y Ito concluded that one half of Japanese industries contributed aggregate TFP in the estimated period. As the reason why IT capital does not contribute greatly to economic growth in this research, Y Ito indicated that the environment for the usage of IT capital has been inadequate in Japan. The result that finance industry contributed to aggregate TFP is worthy of attention. This result bears out the authors' analysis in Chaps. 7-9.[12]

Shinozaki (2001b) carried out the regression analysis across industries using the input-output table for Japan in 1995. Shinozaki constructed the data of industrially classified IT investment and IT labor and regressed IT ratio (IT investment/IT labor), general equipment ratio (general investment/general labor), IT capital equipment ratio (IT investment/general investment),and IT labor ratio (IT labor/general labor) to labor productivities. As a result, Shinozaki concluded that IT quality did not make progress, because IT equipment and IT labor were not statistically significant.

In the meantime, using the fixed capital matrix of the input-output table for the 1975-1995 period in Japan, Kuriyama (2002) constructed the capital stock data of 27 industries, and estimated the Cobb-Douglas production function for IT capital, non-IT capital, and labor in the cross section. He reached the following conclusions:

1. The production function of the non-manufacturing industry exhibited constant return to scale.
2. Although the production function of the manufacturing industry showed decreasing returns to scale, this tendency had weakened for IT capital over time.
3. The elasticity of labor productivity with respect to IT capital in the manufacturing industry is seven times that of the non-manufacturing industry.
4. The increasing rate of IT capital in the manufacturing industry was larger than that of the non-manufacturing industry, and the amount of IT capital in the non-manufacturing industry was larger than that in the manufacturing industry.
5. The estimates of the elasticity of labor productivity with respect to IT capital were significantly positive.

As we have seen, the industry-level research for Japan has just started.

3.5 IT Effect on Employment for the Whole Economy

Minetaki (1986) estimated the cost share function deduced from the production function with production labor, non-production labor, and general capital except for IT, and with IT capital using pool data of 11 Japanese industries in the 1980-1998 period. As the result, there was the substitution relation

---

[12]See Chaps. 8 and 9 for details.

between production labor and IT capital in all industries analyzed. In particular, it was confirmed that the substitution tendency between IT capital and production labor was strong in chemical, electrical machinery, and precision instrument industries. That is to say, IT investment has a negative effect for employment in Japan, as long as the production is fixed.

On the other hand, the research of the Ministry of Internal Affairs and Communications (2001) estimated that IT capital increased the employment of about 2 million persons in Japan in the 1990-1999 period. Job growth due to the demand increase in IT utilization, the demand increase for the IT departments, and the increase in compensation of employees supplemented the decline in employment due to the increases in the labor productivity. It was proven that the increase in employment occurred in all industries, although job growth occurred in the service industry.

Harada and Okamoto (2001) suggested that the stable real wage led to business extension in the USA in the latter half of the 1990s. They indicated that the replacement of unskilled labor by IT allowed wages to remain steady. In Japan, the real wage rate rose for reduced working hours and low inflation by tight monetary policy. They believed that this squeezed firm profits and helped to prolong the economic recession. They considered that reduced working hours and low inflation weakened the business extension effect of IT in Japan.

Shinjo and Zhang (2003) also indicated that IT capital and labor or energy are substitutes for the 1975-1998 period in Japan and the USA. As mentioned above, the effect of IT investment on employment is not confirmed in the analysis of the entire Japanese economy.

3.6 Summary

In the USA, real IT investment consistently increased in the 1990s. As in the studies of Oliner and Sichel (2000) and Jorgenson and Stiroh (2000a, 2000b), the remarkable contribution of IT capital to economic growth would be verified by use of data from the USA in the latter half of the 1990s, in the analysis of growth accounting.[13]

On the contrary, in Japan, real IT investment decreased with the collapse of the bubble economy, and it increased in the latter half of the 1990s.[14] In the analysis of growth accounting of Japan, the contribution and the contribution

---

[13]K Matsumoto (2001) indicated that the high economic growth rate of the USA was generated by the demand factor and asset effect of the stock price rise, and that the economic structure change by IT was non-existent. He warned that it was dangerous to believe the analytical result, because the recent USA studies did not contain the regression period after 2000 in the analysis period.

[14]The method for estimating information investment value from input-output tables was adopted in many studies. For example, see Shinozaki (1999), Shinjo and Zhang (1999), Economic Planning Agency (2000), Toshida and Japan Economic Re-

ratio of IT capital to economic growth rate were drastically different in various studies like Y Ito (2001) and Ministry of Internal Affairs and Communications (2001).

Although the Ministry of Internal Affairs and Communications (2001) estimated marginal productivity from the production function, Y Ito (2001) did not calculate the marginal productivity from the estimated production function. If excess marginal productivity of IT capital exists, and the contribution rate of IT capital is used, the contribution of IT capital to economic growth will be found to be small. The difference between these results of analysis depends on this factor.

Jorgenson and Motohashi (2003) confirmed the contribution of IT capital service to economic growth in Japan on a scale similar to that in the USA in the 1995-2000 period. Their conclusion that the low rate of economic growth in Japan is due to the negative contribution of labor is consistent to that of Harada and Okamoto (2001).

In the meantime, the marginal productivity of IT capital of Japan, which Shinozaki (1998) calculated from the estimate of the macro production function, is bigger than the value of the USA which he calculated in the same way. The under-investment in IT capital influence this result. Shinjo and Zhang (2003) confirmed higher benefit-cost ratio of additional IT investment in Japan than in the USA. It is impossible that the analysis of the production function with IT capital in the USA simply compares with the result of Shinozaki (1998), because it is done at firm-level.

The productivity paradox in the non-manufacturing industries is reported in great numbers in industry-level analyses. In addition, the productivity effect in the manufacturing industry of the USA could not be confirmed in a series of studies by Morrison and Berndt during 1991-2001. Measurement error of the data seems to influence these study results. In studies by Siegel and Griliches that improved the quality of the data, the productivity effect in manufacturing industries was confirmed at industry-level.

In the research by Shinozaki (2001b), the IT effect on labor productivity was not confirmed, and industry data from one period seemed to influence this result. In the analysis in one period, for the IT effect may be estimated to be small, particularly when the IT effect has a time-lag. Although there is the constraint of constructing the data from the input-output table, it is necessary to construct the industry-level data throughout the long term.

Kuriyama (2002) estimated the production function using the cross-section data for 1975, 1980, 1985, 1990, and 1995, for which the input-output tables are utilizable. It was confirmed that the elasticity of the non-manufacturing industry was smaller than that of the manufacturing industry, although the estimated result was statistically significant. IT investment in the non-manufacturing industry was excessive, even if the qualitative factor was ad-

---

search Center (2000), K Matsumoto (2001), Ministry of Internal Affairs and Communications (2001).

justed, and IT effect was small. This result is similar to the study of Matsu-daira (1998), which is reviewed in Chap. 4.

In the USA, statistical data have been collected considering the quality effect. It is impossible to carry out international comparison, because the definition of IT investment in Japan is different from that in the USA. Improvement of the data by an international unification standard and improvements to the estimate method are desirable. In studies of the productivity effect on IT capital for the whole economy and at the industry-level, it is impossible to reach an established view. There is limitation in including qualitative factors, such as the modification of business organization and change of the behavior pattern of the economy, which are the essence of IT revolution in the macro model.

Firm-Level Analysis of Information Systems Investment

S. Watanabe, Y. Ukai, and T. Takemura

4.1 Introduction

By improving IT-related data at the firm-level, research on the economic effects of IT investment was widespread in the United States in the 1990s. For example, the International Data Group (IDG) has handled IT-related data for 500 firms of the manufacturing and service industries since 1987. The effect of IT investment would be confirmed by using these data in the USA. Moreover, the research that has also confirmed the positive firm-level effect in non-manufacturing industries, which cannot have confirmed the positive economic effect of IT investment at the aggregate level, is emerging.

The analytical method explained in this chapter can be classified into two main groups. The first is an estimate of the firm-level production function with IT capital. In this method, it is not necessary to put the neoclassical assumptions on a production function like growth accounting. Various estimates are possible by selecting the Cobb-Douglas production function or trans-log production function, the production factor such as IT capital, non-IT capital, IT labor and non-IT labor, and output on value added or including raw material, when the production function is estimated. This analytical method will be called the production function approach.

The second method is the estimate of the investment function that is obtained from the maximization of the expected discounted value of profits. We define this analysis procedure as the firm value approach, because this analytical method uses the firm value.

By establishing new methods, we are able to independently check the increasing number of research studies on the organization reform effects of IT. In addition, the IT effects on employment and wages at the firm-level are considered in this chapter, following from the examination of the IT effect on employment at the aggregate level in Chap. 3. Table 4.1 summarizes the main research handled in this chapter.

Table 4.1. Firm-level studies on economic effects of IT investment

Method	Studies
Growth accounting	Brynjolfsson and Hitt (2000), Yang and Brynjolfsson (2001)
Production function	Lichtenberg (1995), Brynjolfsson and Hitt (1993, 1995), Hitt and Brynjolfsson (1996), Matsudaira(1998)[J], Prasad and Harker (1997), Ukai and Kitano (2002)[J], Motohashi (2003)[J], Takemura (2003)[J]
Firm value	Brynjolfsson and Yang (1999), Brynjolfsson, Hitt and Yang (2000), Ukai and Watanabe (2001)[J], Ukai and Takemura (2001)[J]
Employment effect	Berndt (1992), Krueger (1993), Brynjolfsson, Malone, Gurbaxani, et al. (1994), Suruga (1991)[J], Obara and Otake (2001)[J]
Organization effect	Brynjolfsson and Hitt (1998), Bresnahan, Brynjolfsson and Hitt (1999), Black and Lynch (2000), Tanaka (2001)[J], Kuriyama (2002)[J], Ukai and Watanabe (2004)[J]
Management strategy effect	Ukai (1997)[J]

[J] Studies on Japan

4.2 Studies on Productivity

By using the Bureau of Labor Statistics (BLS) data for the USA banking in the 1967-1980 period, Brand and Duke (1982) showed that the growth rate of labor productivity in the banking sector per working hour had a yearly average of 2.1% for the 1967-1973 period and this fell to 0.7% for the 1973-1980 period.

The early firm-level studies of the effects of IT investment were not based on econometric models. One of the causes of this is that the firm-level data in the 1980s were poor. However, in 1990s, the important studies on the effect of IT investment using the firm-level data appeared along with the improved data of IDG.

Brynjolfsson and Hitt (1993) estimated the Cobb-Douglas production function with IT capital, non-IT capital, IT labor, and the other production factor using the panel data of IDG for 367 firms, including service industries chosen from 1000 firms reported in the economic magazine "Fortune" in the 1987-1991 period. However, IT expenditure was used instead of IT labor numbers in this estimate so that the data of IT labor only became available in 1990. It is necessary to note that bias may have been generated in the estimated result, because the likelihood of differences in the quality of labor affecting the IT working wage was high. As a result of their analysis, it could be confirmed that the regression coefficient of IT capital was only 13% of the size of the regression coefficient of non-IT capital, although the regression coefficient of IT capital is positive and statistically significant. To the contrary, the regression coefficients of IT labor was over four times that of IT capital. Brynjolfsson and Hitt considered that the productivity paradox had vanished by 1991.

In addition, Brynjolfsson and Hitt (1995) estimated the production function on value added with IT-related capital which totaled IT capital and IT labor, non-IT capital, and non-IT labor. They used IT-related capital because they considered IT capital and IT labor to be complementary. However, it is necessary to make a stock variable of IT labor to make IT-related stock variable by combining IT capital stock and IT labor flow. They assumed that IT labor stock was three times the IT working expenditure, considering that IT labor stock is depreciated over 3 years. This IT-related stock does not include communication equipment, peripheral equipment, and software. However, IT labor stock may be related to software production.

In Brynjolfsson and Hitt (1995), the model of Brynjolfsson and Hitt (1993) was expanded in the following three points: control of characteristics of firms and industries, adoption of the trans-log production function, and division of data (manufacturing and service industries, sectors with measurable outputs and immeasurable outputs, firms in which corporate profits grow in the 1988-1992 period and those showing no growth). By comparing the estimated result of the fixed effects model with the firm characteristics with the estimated result using pool data, they concluded that half of the productivity effect by IT capital could almost be explained by the firm characteristics. Unobserved organization factors might have influenced the productivity effect of IT capital. There was no difference between the estimated result of the trans-log production function and that of the Cobb-Douglas production function. This fact implied that we can verify the IT effect by using the simple Cobb-Douglas production function. In addition, they separately estimated the production functions of manufacturing industries and service industries and examined the null hypothesis that the regression coefficients of IT-related capital were equivalent in both industries. This hypothesis could not be rejected. Moreover, the null hypothesis on the equivalence of the regression coefficients of IT-related capital of the sectors with measurable outputs (manufacturing industries, mining industry, transportation industry, and so on) and the sectors with immeasurable outputs (finance service and other service industries, and so on) was also not rejected. In addition, the regression coefficient of IT-related capital of the firms in which corporate profits grew in the 1988-1992 period was equivalent to that of non-growing firms. They considered the analysis should be carried out by separating the data on the basis of the form of organization and management strategy.

In short, Brynjolfsson and Hitt (1993, 1995) confirmed that the productivity effect of IT capital was statistically significant, even if service industries were contained in the analytical data. In the aggregate level analysis examined in Chap. 3, the exact effect of IT capital could not have been easily confirmed. However, they themselves viewed with skepticism that the regression coefficient of IT capital became equivalent, even if it is estimated by dividing the data into three types.[1] It is necessary to evaluate results of analysis with attention drawn to this point.

Lichtenberg (1995) described research that reinforced the defect of Brynjolfsson and Hitt (1993). He estimated the Cobb-Douglas production function on gross output with four production factors of IT capital, non-IT capital, IT labor, and non-IT labor. In this analysis, two kinds of IT-related data of IDG data ($n = 114$) by "Computerworld" magazine in the 1988-1991 period and data ($n = 388$) by the "Information Week" magazine in the 1990-1993 period were used. Lichtenberg made IT capital and supplemented missing data with strong assumptions. However, only the computer hardware was included for IT capital in his research. He estimated the production function by using two kinds of data independently or the mean value of two data. By putting strong assumptions between firm profits and value added production functions, he carried out the analysis because the data of intermediate input could not be used.

Using the estimated result, Lichtenberg tested whether the ratio of marginal productivity of IT capital to marginal productivity of non-IT capital is equal to the ratio of each user cost. This implied whether the excess marginal productivity of IT capital to non-IT capital exists. It is possible to transform the former ratio to the ratio of the regression coefficient of IT capital to that of non-IT capital.[2] In this test, it is possible to test the possibility in which excess marginal productivity by IT capital is arising, even if the regression coefficient of IT capital is considerably smaller than that of non-IT capital. Lichtenberg (1995) showed that IT capital and IT labor had excess marginal productivity in most estimation. This fact means that large increases of output and productivity are produced by slight increases in IT expenditure. Lichtenberg also examined marginal rates of substitution between non-IT labor and IT labor. As the result, it could be confirmed that 6 non-IT workers could be substituted with one IT labor for a fixed output.

Unlike studies of the USA, which is blessed with data of IT capital, studies in Europe have not been easy. For example, Kwon and Stoneman (1995)

---

[1] See Brynjolfsson and Hitt (1995).

[2] Brynjolfsson and Hitt (1993) compared the regression coefficient of IT capital with the regression coefficient of non-IT capital. However, they did not test the excess marginal productivity.

estimated the Cobb-Douglas production function with capital, labor, and technology using the data of 217 firms in manufacturing industries in the UK in the 1981-1990 period. This study was analyzed by dividing the technology into five factors including computers, because IT capital and IT labor data could not be obtained as in the research of the USA. Kwon and Stoneman carried out simultaneous determinations of the output and production factors, and the endogenous nature of the technology. Simultaneity bias occurs in the estimation when production factors and the endogenous nature of the technology are disregarded. Concretely, capital and labor required from the first-order condition of the profit maximization were substituted for capital and labor of the production function. The endogenous nature of the technology was determined using the conditional expression on adoption point of time of the technology on the profit maximization of the firm. Five technology data have been adopted from the investigation of Newcastle University for 1981, 1986 and 1993. In the estimation, a two-step method was used in which the first estimated result of the profit maximization of the firm was substituted into the production function. The result showed that computer capital and NC (numerical control) machine capital within five kinds of technology adoption had positive effects on productivity and output.

Thus, a positive productivity effect on IT capital in many estimated results of the production function with separation of IT-related production factors and decomposition of the technology was shown. However, these study results are not limited to specific industries. The positive effect of IT investment might be shown for the whole economy according to the data of other industries, even if it is ineffective for IT investment in some industries. On this point, the analysis limited to the banking of the USA is worthy of attention.

Prasad and Harker (1997) analyzed the production function of banking using the IT expenditure of 47 banks obtained by questionnaire on human capital management and management of the USA banking for the 1993-1995 period.[3] The analyzed banks had assets exceeding 6 billion dollars. It was discovered that the elasticity of production with respect to IT labor was 0.19 or 0.25. It was statistically significant, although the elasticity of production with respect to IT capital is not significant by the estimate that used the total loans and deposits or net income as output.

In this research, it is necessary to note that the precise estimate of the production function of service industries is difficult. This is because the definition of the product of banking is uncertain. They used total loans and deposits, net income, ROE (return on equity), and ROA (return on assets) as outputs of banking in the USA. However, it is natural that numerical indexes that aggregate each component of the operation assets using the Divisia index are regarded as a product, when the bank offers multiple services in which the

---

[3]Maimbo and Pervan (2002) reviewed the papers that analyzed the effect of IT capital in banking.

quality change is intense.[4] However, the result of the analysis, in which elasticity of production with respect to IT labor is statistically significant, is similar to that of Brynjolfsson and Hitt (1993).

The trend of the new research of the USA is the analysis of the productivity effect at firm-level, considering the time-lag effect of IT investment and the existence of intangible assets correlated with IT capital. For example, Brynjolfsson and Hitt (2001) analyzed the relationship between computers and productivity growth rate using balanced panel data of 600 firms from "Fortune 1000" of the USA in the 1987-1994 period. They estimated the elasticity of production with respect to IT capital in the production function from growth accounting, and calculated the marginal productivity of the computer using it. It was shown that the marginal productivity of the computer was two to five times larger when using a 7-year difference of IT capital than that obtained when using a 1-year difference of IT capital. This result implied that the computer contributed to long-term productivity growth. Therefore, it is possible to understand that the computer is bringing about technological change and organization reform in the long term. Moreover, Yang and Brynjolfsson (2000) analyzed growth accounting with intangible assets concerning IT. They found that the contribution of intangible assets was larger than the contribution of IT assets, and the growth rate averaged 1% per year in the 1990s.

In the above two studies, the analysis of not only short-term effects of IT investment but also comprehensive long-term effects of IT investment were possible. In the future, mainstream research on productivity might advance in this direction.

With regard to studies on Japan, Matsudaira (1998) performed an analysis of the production function with IT capital. This study was based on the technique of Brynjolfsson and Hitt (1995), and Lichtenberg (1995). Using the original investigation data on IT assets and expenditure in 1997, Matsudaira analyzed the effect of IT investment on 228 Japanese firms. However, the finance industry was excluded from the analysis. The IT data were based on questionnaires concerning the acquisition cost of present computer hardware and network equipment. In particular, the approximate market price of all computer hardware assets used in each firm (computer hardware, computer software, outsourcing cost, information system personnel expenses, miscellaneous expenses) was asked. Although there is merit in simply obtaining the

---

[4]Takeda (1998) used a numerical index that totaled each component of the operation assets by the Divisia index to reflect the composition of the product and change of the product price in the estimation of the productivity of regional banks. The composition of the product must be fixed when the sum total of the operation assets is used for the quantity index of the product. That the bank offers multiple services in which qualitative change is enormous also made the measurement of an output difficult. See appendix II of Kasuya (1989) about the numerical value index by Divisia.

stock amount without considering depreciation, this method introduces inaccuracies.

Combining the financial data of 228 firms (150 firms in manufacturing industries and 78 firms in non-manufacturing industries) with the questionnaire data, Matsudaira analyzed the economic effects of IT, especially the marginal productivity of IT investment in Japan. The result of the analysis showed that the regression coefficients of IT capital were significantly positive in all industries and the elasticity of production and marginal productivity were positive. The analysis of manufacturing industries showed that the marginal productivity of IT capital was positive and about 10% larger than that of other capitals.

In the meantime, the regression coefficient of IT capital was small and not significant in non-manufacturing industries. In short, it implied that marginal productivity of IT capital in non-manufacturing industries was zero. This is in contrast to the study of Brynjolfsson and Hitt (1995) in which the productivity effect of IT capital was confirmed, even in service industries. Matsudaira regarded the marginal productivity of IT labor as zero, because the regression coefficient of IT labor in both manufacturing industries and non-manufacturing industries are not significant. This is in contrast to the study of Lichtenberg (1995), which evaluated the productivity effect of IT labor in manufacturing industries of the USA.

Matsudaira indicated the firm characteristics that might be correlated with IT capital and had not been introduced into the production function might enlarge the regression coefficient of IT capital. An analysis was also carried out in which the research and development (R&D) capital was introduced into the production function as the firm characteristics that might be correlated with IT capital. As the result, it was found that the regression coefficient of IT capital was slightly lower.

The analysis of Matsudaira in Japan showed that IT capital is not effective in service industries and IT labor is not effective in all industries. This result is the reverse of that found in the USA. Although it is conceded that imperfections probably exist in the IT capital data, given that the analysis was carried out at a point in 1997, the observed effect may well be small. Therefore, analysis using panel data is desirable.

Ukai and Kitano (2002) estimated the Cobb-Douglas production function of Japanese banking in the 1996-2001 period using the data of management reconstruction plans, which were submitted to the Financial Services Agency. They carried out the cross-section analysis for every single year. Statistically significant results were found in their analysis in the 1998-2000 period. It has been confirmed that the elasticity of production with respect to IT capital tended to increase, although the elasticity of production with respect to non-IT capital became lower over time.

Takemura (2003) examined the impact of information system on productivity and efficiency in the Japanese banking industry by using data set at the firm-level[5]. Takemura (2003) estimated the Cobb-Douglas production function with information system asset and/or computer software asset by using a stochastic frontier model as an econometric tool, and compared with Prasad and Harker (1997), who analyzed banking industry in the USA. Takemura found that additional investment in information system assets, especially computer software assets, made positive contribution to output of banking (the sum of loan and deposits excluding bad debt) in the 1993-1997 period (the earlier age of the post-third-generation on-line system), but might makes no contribution in the 1998-1999 period. In addition, he found that banking industry has higher efficiency than the other industries in Japan.

Motohashi(2003) estimated the effects of IT investment and information network use (13 types of application such as intra firm network, production , and human resource management) to the value added from the estimates of first difference of the Cobb-Douglas production function. The dataset of Ministry of Economy, Trade and Industry (METI) was used in this analysis. This dataset includes the application type of information network use, organizational characteristics, and performance variables of manufacturing, wholesale, and retail firms in 1991, 1994, 1997, and 2000.The result showed that the network effect to productivity differed on the type of its application and firms which had inter firm network and collaborative activities with other firms such as joint production and joint R&D except for outsourcing of production. had bigger IT effect on output than either of them.

The estimate of the production function of Japan at the firm-level is not usually carried out in this fashion. This is in contrast to the analysis at the aggregate level, which calculates IT investment value using the input-output table by Shinozaki (1996, 1998, 2001a), and is widely conducted in government agencies. There appears to be some reluctance to sufficiently open IT-related data at the firm-level to the public.

Because it is difficult to statistically aggregate the service trade like the analysis of banking in this section, the measurement problem of the non-manufacturing industries is more serious than that of the manufacturing industries. To avoid the problem of measurement error, a new research trend using different indexes, such as the stock market value, is becoming more common.

4.3 Contribution of IT Capital to Firm Value

Brynjolfsson and Yang (1997) used the firm value as the alternative to productivity as the index that measures the economic effect of IT.[6] The model in their studies is the reduced form of investment function using the technique of the dynamics optimization. The condition in which marginal productivity and

---

[5]Data set is the same as one in Subsect. 9.2.2.

[6]Subsect. 8.2.1 gives explanation of Brynjolfsson and Yang (1999) for details.

capital user cost become equal under the assumption of profit maximization will be obtained when the capital equipment is used as a production factor. This condition implies the equality of increase in the future profit and investment goods price by the investment. It then becomes important how the prediction for the future is dealt with in the model.

Tobin's q theory is a suitable method for estimating the investment function without assuming the prediction formation clear. Tobin (1969) indicated that investment is the increasing function of the ratio (average q), which the numerator is a firm's market value and the denominator is a firm's book value of the firm assets. In short, the work in estimating the complicated investment function was simplified by replacing the prediction variable in the stock market with the market value of the firm. Future profit expectation is correctly evaluated in the stock market, if the perfect competition of the capital market is assumed.

Hayashi (1982) clarified the condition that average q is equal to marginal q. Marginal q is the ratio of the change in the value of the firm to the added capital cost.

This condition is to assume homogeneous function on both the production function and the adjustment cost function. However, there is a criticism that the condition of Tobin's average q being equivalent to the marginal q is unrealistic. There are also other criticisms that (1) diversity of the financing method of the firm and the change of production technology lead to the computational complexity of the assets re-acquisition cost and (2) the model can not treat stock price fluctuations in short run.[7]

Brynjolfsson and Yang (1997) deduced the model that explains the firm value (stock value plus debt) by IT assets using Tobin's average q. The numerator of Tobin's average q is the firm value (explained variable in this model) and the denominator is the assets (explanatory variables in this model). Therefore, they determined which of the assets raised the firm value more by comparing the size of the parameter of each asset.

Brynjolfsson and Yang (1997) indicated that the computer may form intangible fixed assets affecting organization and human capital.[8] The reacquisition cost of the assets, which is a denominator of the average q, has been

---

[7]Many researchers of economics and management use the market value for firm evaluation. In particular, the research technique of "event study" is widely used in the USA. This technique analyzes the effects of events on stock prices to determine how various events, which may be relate to the economy or management issues, affect the firm evaluation. A technique that estimates stock market value of R&D capital has also been used by Griliches (1981). Hall (1993) has also investigated the connection between R&D and market value.

[8]Ogawa and Kitasaka (1998) analyzed whether the "bubble economy" really existed in the capital market in Japan using Tobin's q theory. They estimated a macro marginal q and an average q in Japan and confirmed the existence of the bubble because these two values separated in the latter half of the 1980s.

less evaluated when all assets concerning the production of the firm are not informed.

The coefficient of the information assets can easily be increased above 1 when this point is considered. The assets realized by on-the-job training (OJT) costs and restructuring costs on corporate structure and custom for dealing with new technology and the change of a demand did not appear on the balance sheet exactly. In addition, assets such as personal computers should be processed as a cost in the accounting. The computer assets should be handled as assets that substantially affect the production, because the computer is used for a certain time. It is a technique in accounting whether the firm deals with the computer as a cost before use or as depreciation cost in use.

Brynjolfsson and Yang (1997) estimated this firm value model using the data of non-finance industries in the 1987-1994 period. As a result of the analysis, the regression coefficient of IT assets was about 10. This implied that an increase in IT assets of $1 raises the firm value by about $10. This regression coefficient is larger than the regression coefficients of other assets. Furthermore, studies that introduced intangible assets explicitly into the model have emerged. Brynjolfsson, Hitt and Yang (2000, 2002) made a kind of index by analyzing the results of questionnaires for the firms on principal component analysis. They introduced the index as the proxy variable of intangible assets into the firm value model. As the result, although the regression coefficient of IT assets was lowered to about 5, the regression coefficient of IT assets was still larger than those of other assets. They also indicated that the effect of $1 of IT capital on the market value of firms with decentralized organization structure was $2-$5 higher than the effect of IT capital on the market value of firms with centralized organization structures.

Ukai and Watanabe (2001) analyzed the effect of information system assets of Japanese banking on the market value using the data of information system investment collected by their questionnaire in 1995, 1996, and 1998. The information system investment was defined as the system development cost and the outsourcing cost of software and hardware. The assets were estimated from these costs. It was shown that the information system assets greatly contributed to the stock market value in comparison with other assets, which was similar to the result of analysis of non-finance industries by Brynjolfsson and Yang (1997). Ukai and Takemura (2001) analyzed how much the software assets contributed to the stock market value. These assets were derived from financial statements of listed firms in the first and second sections of Tokyo and Osaka stock exchanges in Japan in the 1993-1999 period. In this study, the fixed effects models were adopted in all estimated results. This indicates the possibility that intangible assets, which correlate with information system assets, exist in Japanese firms, in a similar manner Brynjolfsson and Yang (1997) and Brynjolfsson, Hitt and Yang (2000) emphasized the existence of intangible assets for information system assets for the USA firms. These findings are discussed in Chaps. 8 and 9. Thus, although the object area and targeted companies are different, the unified view is that IT assets have positive effects for the stock market value of the firm, and this has been obtained in the firm value approach.

4.4 Analysis of Organization Effect and Management Strategy Effect

In studies of organization effects of IT capital, there appear to have been studies on the organization reform effect and the organization scale effect. To begin with, we review the studies on the organization reform effect.

Brynjolfsson and Hitt (1998) were the first to show attention in the organization reform effect by conducting a questionnaire survey on business organization by telephone. The sample size was 379 firms. Brynjolfsson and Hitt analyzed the organization effect on IT investment demand and productivity of IT investment using these data. The analysis showed that IT investment demand was high in organizations that had adopted decentralized organization, especially for self-management teams, and in organizations that strongly invested human capital and training. The analysis showed that the productivity effect of IT investment was higher when the organization had these tendencies.

Black and Lynch (2000) analyzed the relationship between workshop innovation and productivity or wages. The result showed a positive relationship between the ratio of non-managers using the computer and productivity. It found that firms with workshop innovation showed high productivity and had high wages.

In addition, Bresnahan, Brynjolfsson and Hitt (1999, 2002) analyzed whether the manager who carried out IT investment accomplished mutually complementary reform for the organization and human capital. In this study, the survey data for the large firms that handled the organization and human capital were drawn from the earlier work of Brynjolfsson and Hitt (1998). The analytical result on production functions and IT demand function showed that IT capital gave wide responsibility to the worker, decentralized decision making, and had a complementary relationship with the organization with the self-management team.[9] In actual fact, the elasticity of production with respect to IT investments on the firms with the decentralized organization

---

[9]Jorgenson and Fraumeni(1992a) estimated the contribution of human capital investment to USA economic growth at the aggregate level. They used the output of the education sector as an investment in education of the noneducation sector. The investment in human capital leads to the substitution between more effective workers and less effective workers. The substitution was described as the growth in labor quality in the growth accounting. Their analytical result showed that the contribution of labor input was bigger than that of capital input and labor quality accounted for more than two fifth of the contribution of labor input in the 1948-1986 period. See Jorgenson and Fraumeni (1989, 1992a, 1992b) for detail.

was 13% larger than that of the average firms. The fact that IT capital and organization are complementary with the skilled worker was verified.

The Economic Planning Agency Research Bureau (2000) issued a questionnaire to listed firms in the first and second sections of Tokyo and Osaka Stock Exchanges and non-listed firms extracted from the firm data of Teikoku Databank using the analysis procedure of Bresnahan, Brynjolfsson and Hitt (1999). They analyzed the data of 482 firms. They made indexes of IT, human capital, and organization from the questionnaire and calculated Spearman partial rank correlation coefficients controlling industry, firm scale, and so on. They made a group index by combining two indexes of IT, human capital, and organization and analyzed the effect of every group index on total factor productivity. These analyses provided the complementary relationship between IT and human capital or between IT and organization, as well as the result of the USA firms.

Similarly, Kuriyama (2002) issued a questionnaire to listed firms in the first and second sections of Tokyo and Osaka Stock Exchanges and used the data (110 firms, 7.94% response rate) to estimate the Cobb-Douglas production function. Although the effects of the personnel system and IT on the productivity in Japan was similar to that in the USA, Kuriyama concluded that the centralized organization had higher productivity than the decentralized organization. This is not consistent with the result of Bresnahan, Brynjolfsson and Hitt (1999) and the Economic Planning Agency Research Bureau (2000) that a increase in human capital, decentralization of decision making, and IT boost the productivity. It is generally considered that USA firms are characterized by centralized decision making and decentralization of the personnel evaluation, while Japanese firms are characterized by decentralized decision making and centralization of personnel evaluation. Kuriyama (2002) considered that USA firms prevent excessive centralization of decision making by decentralization of the organization and delegation of authority, while, on the other hand, centralization of Japanese organization has adversely improved productivity since decentralization of organizations has advanced in Japanese firms.

There is a complementary relationship between the view that organization reform raises the productivity and the view that IT raises the productivity. Black and Lynch (2000) and Bresnahan, Brynjolfsson and Hitt (1999, 2002) confirmed this effect in the USA. The analysis by the Economic Planning Agency Research Bureau (2000) in Japan is also consistent with this relationship. However, the conclusion of Kuriyama (2002) is in contrast to the result of the Economic Planning Agency Research Bureau (2000) and Bresnahan, Brynjolfsson and Hitt (1999, 2002). Kuriyama concluded that concentration of decision making has an effect on the productivity of IT. However, there is a strong possibility that sample selection bias occurred in the work of Kuriyama, because the data are not extensive.

Ukai and Watanabe (2004) analyzed the relationship between IT capital and human capital or the organization using the questionnaires for the post masters in Kinki area, Japan . Ukai and Watanabe calculated the rank correlation coefficient and the cross-tab using the data of 326 post offices. The analytical result showed that the complementary relationship between IT capital and the organization or human capital would lead to increase in the amount of postal savings per post-office worker.

Let us leave the organization reform effect and turn to the organization reform scale effect. Brynjolfsson, Malone, Gurbaxani, et al. (1994) investigated whether IT investment in the USA reduced the firm scale (the number of workers). They used IT investment data and indexes, such as the employee number per firm, in the 1976-1989 period. The result of their analysis showed that IT investment reduced the average firm scale with a time lag of 2-3 years. In addition, Hitt (1999) developed an index that represents the degree of vertical integration of the firm, and analyzed the relationship between the index and IT investment. His result of analysis showed that IT investment softened the degree of vertical structure in the firm.

In the meantime, Tanaka (2001) analyzed the effect of IT investment on the firm scale in Japan. He first carried out factor analyses on all contributions of IT investment to the productivity. He concluded the reason that IT investment increased the labor productivity (value-added/the number of workers) was the increase in sales, which surpassed the increase in the transaction cost. It is generally considered that increases in the firm scale (the number of workers) increase the transaction cost in the organization and decreases in the firm scale increase the transaction cost in the market.

Tanaka used panel data of 479 Japanese firms of "corporate activities basis investigation" and "information processing field study" by Ministry of Economy, Trade and Industry in the 1995-1997 period. IT capital stock for this analysis included hardware and software. Tanaka (2001) confirmed that IT investment is complementary with the number of workers. However, Tanaka found that package software brought out standardization and reduced employees. This result is consistent with study results of IT effects on employment.

Using the operating cost as a proxy variable of transaction cost, Tanaka obtained the result that IT investment increased the transaction cost. In addition, he concluded that IT investment increases sales, which surpass increases in the transaction cost. This result is quite a contrast to the conclusion of Hitt and Brynjolfsson (1996), examined in Chap. 3, that IT did not increase the firm profit. However, it is necessary to keep in mind the differences in the countries, object industries, and estimation periods between these studies.

However, it should be noted that using the operating cost as the transaction cost could be a problem because the operating cost includes all costs except the transaction cost. This analysis showed only the effect that IT investment reduces employment.

For the analysis of the firm scale, it is necessary to find a proxy variable that accurately represents the transaction cost or directly analyzes the organization structure as shown by Hitt (1999).

Finally, let us consider the management strategy effect on IT investment. Ukai (1997) performed a factorial analysis on the relationship between information system investment per employee in the system development sections of Japanese banks. He collected data from 27 banks concerning information system investment and management strategies for 1995 by mailing out questionnaires. Three management data groups and investment rate were analyzed by factorial approach. Two groups were rejected at a 95% level; however, the relationship between central bank or government and the management were significant at the 95% level. It was found that the more independent type of management to the central bank tended to make more investment than the less independent type of management. The authors call this the management strategy effect on IT investment.

4.5 Effect on Working Conditions

The effect discussed here can be classified into the IT effect on employment and the IT effect on wages. Berndt (1992) concluded that the increase of IT capital reduced the number of blue-collar workers, and, on average, increased the number of white collar workers. The labor productivity per capita will become lower if computer investment heightens the productivity and drastically increases the number of information workers.

Suruga (1991) analyzed the IT effect of Japanese banking on employment in the second-generation on-line system. He estimated the employment function of 58 regional banks with data in 1987 that included: (1) sum of depreciation costs of movable assets and computers, and OA related rental lease fees, (2) memory capacity of general-purpose computers, and (3) the proportion of automatic teller machines (ATMs) in the total number of cash dispensers (CDs) and ATMs combined. These data types were used as measures of computerization and office automation. As result, Suruga used (1) sum of total loan and deposits, bonds and deposit certificates, and (2) net income.

The only statistically significant data of computerization and office automation were the proportion of ATMs to the total number of CDs and ATMs. He concluded that computerization and office automation tended to replace clerical staff. This seems to have been the effect of the second-generation on-line system discussed in Chap. 1.

In the meantime, a survey conducted by the Ministry of Health, Labor, and Welfare (2001) showed that 60% of the firms that replied did not have an IT effect of employment reduction. However, there were quite a lot of firms with the reduction effect, and some firms considered that the reduction effect would emerge in the future.

Job losses were incurred in sections such as personnel and labor affairs, accounting and financial affairs, and general administration and publicity. In these sections, the amount of "typical work" decreased. As the method of attrition, Japanese firms mainly used non-replacement of retirees, job displacement, temporary transfer, and permanent transfer. The more extreme measures of voluntary retirement and dismissal were used infrequently.

Thus, IT replaces blue-collar workers and supplements white-collar workers, and also replaces white collar workers when IT develops. The study of Osterman (1986) discussed this adjustment process as it occurred in the 1980s. Brynjolfsson, Malone, Gurbaxani, et al. (1994) confirmed the substitution relation between employment and IT in the 1990s. The substitution relation between employment and IT was also shown in the analyses of Suruga (1991) and Minetaki (2001) on Japan. Because Suruga (1991) analyzed data from the 1980s, Japanese banking may have made progress on IT earlier than other industries within that time frame.

With regard to the effect of IT on wages, Krueger (1993) showed that the wage of workers who use computers was about 10%-15% higher than that of workers who did not use computers.[10] It was also reported that highly educated workers used the computer very often and half of increase in the wage premium on education was associated with the use of computer in the 1984-1989 period. However, studies that reject this finding also exist. DiNardo and Pischke (1997) similarly confirmed the effect of the computer on wages like Krueger (1993) by using data from Germany. However, they also noted that simple office supplies, such as pens, also had a weak effect on wages. They understood the relation between wages and computers to be indirect. Unobserved differences in ability that correlated with computer use might also bring out the wage differential.

If this ability is heightened by training as Krueger (1993) considered, wage differential might be eliminated by training. However, training on the computer will not lessen the wage differential if this ability is fixed over time. Obara and Otake (2001) performed similar research in Japan. They measured the effect of the computer on wages by controlling the unobserved differences in ability. The research data of Osaka prefecture on wages and computer use before and after job changes of the same individual was used.

Obara and Otake found the effect of computer use on the increase in wages in level estimation at one period. However, this effect became small and was not statistically significant in their study, when the difference estimation was carried out by the attribution control. Obara and Otake indicated that computer use heightened the wage when better-educated male office workers aged under 35 years used the computer. The result of their study showed that the effect of computer use on the wage differential in Japan might be restrictive. It is necessary to carry out further analyses on whether the ability that correlated with the use of the computer is controllable by training. Thus,

---

[10]The term "complementary relationship" can be defined as the relation between these production factors in which an upsurge of the production in increasing one production factor occurs when the other production factor increase.

research on the wage premium by computer use is developing, and a clear conclusion has not yet been obtained for Japan-USA studies.

4.6 Summary

In this chapter, the production function and firm value approaches were reviewed. Of particular note, various analyses can be performed on the production function approach according to how the production factors are divided and what is included in the definition of IT capital. Studies of the estimate of the production function which divided IT labor and IT capital and those which combined IT capital and IT labor should also be noted. As Brynjolfsson and Hitt (1995) considered, if IT labor and IT capital are complementary and cannot be treated separately, analysis using IT-related capital will be necessary.[11] In this book, the authors use the term "Information System Asset I" to refer to the sum of software, hardware, and information system-related human resources.[12] Analysis using the information system assets based on this definition is discussed in Chap. 8.

Brynjolfsson and Hitt (1995) concluded that there is no difference between manufacturing industries and service industries with regard to the effect of IT capital in the USA. Matsudaira (1998) showed the elasticity of production with respect to IT capital of non-manufacturing industries is not statistically significant and contrasts to the case of the manufacturing industries for which the relation is statistically significant in Japan. It is not possible to make the conclusion that this result is based on data integrity of non-manufacturing industries or that IT capital does not really have an effect in non-manufacturing industries in the period concerned. In addition, although Ukai and Kitano (2002) had statistically significant estimates of the production function in Japanese banking, the significant estimated period was short.

In the meantime, analysis of the effect of IT assets on the stock market value was mainly carried out by the Brynjolfsson group as an analytical procedure that was not based on production function. Brynjolfsson and Yang (1997) and Brynjolfsson, Hitt and Yang (2000, 2002) confirmed that the effect of IT assets was stronger than those of other assets. Ukai and Watanabe (2001) and Ukai and Takemura (2001) obtained similar results.

The research of the Brynjolfsson group has centered on the analysis of the non-financial industries of the USA, while the research of the Ukai group has focused on Japanese banking. Therefore, it is desirable that analyses be conducted on the financial industries of the USA and on the non-financial industries of Japan to allow interindustrial and international comparisons.

---

[11] See Brynjolfsson and Hitt (1993, 1995), and Lichtenberg (1995).

[12] Shimizu (1998) and Ishihara (2000) reviewed papers of econometric analysis on the wage differential. Otake (2001) comprehensively reviewed the effect of IT on employment from the viewpoint of wage differential, employment placement, and women and aged persons.

Studies like Brynjolfsson, Hitt and Yang (2000) to introduce IT capital and intangible assets such as human capital and organization reform into the model have tended to increase. In particular, the Economic Planning Agency Research Bureau (2000) and Kuriyama (2002) performed similar research on Japan using the analysis procedure of Bresnahan, Brynjolfsson and Hitt (1999, 2002). It is interesting to note that the conclusions drawn from these studies concerning the centralization of decision making were directly opposed. Motohashi (2003) also confirmed the importance of collaborative activities with other firms such as joint production and joint R&D in the context of information network. It is necessary to collect data that represent industries and the whole economy to test whether these result are representative of Japanese organizations. Diversified firm-level data are necessary for the analysis of the organization reform effect. The analysis from both economic and management perspectives must be carried out to measure the comprehensive effect of information system investment.

In the USA, an IT capital database has been constructed at firm-level. However, the IT capital database of the USA that Brynjolfsson and Yang (1997) and others used did not contain software data.

In the meantime, a comprehensive database of IT capital in Japan did not exist by 2000. Therefore, questionnaire data were used in Chap. 8 of this book. In addition, the data used in Chap. 9 were originally constructed from financial statements. These comprehensive databases must be constructed in order to deepen the research on the economic effects of IT capital in Japan.

Part III

Positive Analyses of Information System Investment in the Banking Industry

Outlook and Study Process of Questionnaires

Y. Ukai, and H. Nagaoka

5.1 Background and Study Process of Questionnaire Surveys

The number of Japanese nationwide banks contracted from 11 groups in April 1994 to 5 in April 2004. This number is expected to diminish further to 4 by the end of 2005. Among three long-term credit banks, two banks were delisted by the stock exchange markets in Tokyo and Osaka in 1998, and the other bank merged with one of the nationwide banking groups. Even though the number of regional banks is constant at 64, the number of second regional banking organizations decreased from 68 in April 1994 to 50 in April 2004.[1]

In addition, the regulatory agencies of several types of banks that were administered by the banking bureau at the Ministry of Finance from April 1994 became the Financial Supervisory Agency in 1998 and then the Financial Service Agency in 2000.

The Workshop of Information System Investment (WISI) was established by three economists and two computer scientists in the Kansai area of Japan (Kyoto, Osaka, and Kobe) in September 1994, and has conducted research and analysis of information system investment in banking industry during the stormy conditions of the Japanese economy. Two of the authors of this book (Ukai and Watanabe), are confounders of WISI.

The workshop paid particular attention to the White Papers on Financial Information Systems, which have been collected annually by the Center for Financial Industry Information Systems in Tokyo since 1986. However, after careful analysis, it became apparent that all of the public data in the white

---

[1] See the Center for Financial Industry Information Systems (1994) p501 and the Center for Financial Industry Information Systems (2004b) p272.

papers were aggregated data. Therefore, it was clearly impossible to analyze the information system investment effects on individual firm performance.

On the other hand, econometrics entered a new stage in the 1980s without the need for cross-section analysis or time series analysis, and with the ability to conduct panel data analysis for regression and statistical tests based on data of individual firms and households over given periods. The authors contributed to this science with the first panel data analyses of the financial industry in February 1995.

Ukai and Watanabe planned to estimate the market value of information system investment from individual data of equipment investment collected by the Bank of Japan. However, a joint research effort with the Bank of Japan did not eventuate.

Attempts were then made to aggregate scattered data of information system investment from the financial statements to the Ministry of Finance all over Japan, by collating data on computer hardware, software, and systems engineers' wages. It was regrettable that most of the important computer software data had been described in very arbitrary terms in the financial statement of each bank at that time and this situation is explained more precisely in Chap. 6. Still worse was that the personnel costs were described in total and the details of each department were not disclosed at all.

In short, the disclosed financial statements did not fit in well with initial research objectives of determining the values of computer hardware, peripheral equipment, and computer software, and combining them with personnel costs of computer engineers. (a complete estimate of information system investment would combine physical and human resource values) As a last measure, questionnaires were sent out to banks to obtain more precise data that suited the goals of the research.

On the other hand, the financial statements since 1999 have gradually disclosed more details of computer software assets, as precisely explained in Chap. 6.[2] In addition, some private banks were nationalized or put under state control by new laws in 1998, and therefore, public money invested in private banks because of their capital enforcement in 1999. Afterward, the cost items related to mechanization, including several figures related to information system investment, have gradually been submitted to the Ministry of Finance by banks under public control that are required to disclose management reconstruction plans. For the other banks that were not controlled, the figures remain unclear.

Under the conditions described above, Ukai directed students at the Graduate School of Informatics, Kansai University, to coordinate the mailing of the first questionnaire to the 120 banks which had been listed on the Tokyo and Osaka stock exchanges in February 1995. Twenty-six banks returned the answered questionnaires.

---

[2]See Sect. 6.3 and Subsects. 6.2.1 and 6.5.2.

In July 1995, Ukai visited Dr. Daniel E. Sichel, who was then a research associate at the Brookings Institution in the United States and now is a senior economist at the Federal Reserve Board, to discourse with him about the outlook for the answered questionnaires. At that time, Dr. Sichel was writing Sichel (1997), and commented that the estimation of computer investment by using the answered questionnaires would be the first such research in the banking industry in the world. However, this comment later proved to be untrue because the Financial Institution Center at the Wharton School, University of Pennsylvania, had issued research questionnaires to American banks from 1993 to 1995.[3]

Being motivated by Dr. Sichel's comments, the authors hosted the 5th Informatics Forum at the Kansai University Senriyama Campus in October 1995, where Ukai presented the statistical results of first series of questionnaires. Discussions were held with two invited delegates, Mitsuru Iwamura, then the director at the Second Research Unit of the Institute for Monetary and Economic Studies, Bank of Japan and now a professor at Waseda University, and Hisao Nagaoka, then an executive director of Daiwa Bank, and now a research associate at the Research Center of Socionetwork Strategies, Kansai University, and more than 30 computer and financial specialists in the banking industry. This forum received good feedback from these scholars and specialists and a summary of above-mentioned discussion was reported the next day in the Nikkan Kogyo Shimbun (the Daily Engineering and Construction News), Tokyo.[4]

Being pushed by this social acceptance, Ukai and Watanabe mailed a second questionnaire to 120 banks that had been listed on the Tokyo and Osaka stock exchanges in February 1996, based on the statistical analysis of the first questionnaire. Thirty banks returned the answered questionnaires. On the last page of the questionnaire, respondents were offered the opportunity to express an opinion about the ambiguity between operation and investment costs, which arose as an issue after analysis of the first questionnaire. The reactions of respondents to this question were half pros and half cons, and it was definitely helpful for the authors to have a practical sense of information system investment and operation.

The increased number of replies in the second questionnaire was reassurance for the acceptance of the Grant-in-Aid for Scientific Research: Positive Analyses of Information System Investment in Money and Banking Industry 1997-2000, headed by Yasuharu Ukai. In addition to Ukai, the project team consisted of Shinji Watanabe, a research assistant at Osaka Prefecture University, and Hiroaki Aoki, an associate professor at Hannan University. They are all the members of WISI established in 1994.

---

[3]See Prasad and Harker (1997).

[4]See the Nikkan Kogyo Shimbun (the Daily Engineering and Construction News) [J], Tokyo, Japan, October 4, 1995, p21.

The presentation made by Ukai (1997) at the 16th SAS User Conference Japan in September 1997, at Yebisu Garden Place, received useful advice and comments from more than 50 financial and computer specialists in Tokyo metropolitan area.

In October 1997, Ukai reported his revised version of Ukai (1997) at the Bureau of Computer and Information (Bureau of Information Systems from 1998), Bank of Japan, and received comments from Katsuhiro Endo, the director of the bureau, and Hajime Mizuno, the chief of the systems development unit at the bureau. The most important comment was that the personnel cost of information systems tended to be estimated less than the real personnel cost because of the distribution of staff to the branches. However, this comment was largely refuted by responses in the third questionnaire in December 1997.

In October 1997, Ukai set up the first open forum of WISI at the Tokyo Station Hotel and reported the second revision of Ukai (1997) to have a discussion with the information system specialists of the Ministry of Finance, the Bank of Tokyo-Mitsubishi, and Fuji Bank (united with two other banks as the Mizuho Financial Group in 2002). During this meeting, some specialists commented that the wage differences between system development department staff and system operation department staff identified in past questionnaires was derived from the different average ages of staff in those departments. In addition to this comment, the main question that the economy of scale could not be effective in the information system assets, raised by Ukai (1997), was attacked by the opposite opinion that accounting systems could enjoy economies of scale based on practical experience. This experience is derived from the time of the first-generation on-line system through to the third-generation on-line system.

From the lessons of the first open forum, the authors focused on not only questionnaires but also on interviews with information system officials of each bank. Figures in the financial statements and questionnaire answers required suggestions and opinions of these specialist staff members to confirm the data obtained. They had interviews with the officers of the information system departments of all nationwide banks (the so-called city banks) except for Hokkaido Takushoku Bank. Interviews were also conducted with the officers of some regional banks which appeared to be investors in innovative information system technology from their answers to the questionnaires.

In December 1997, Ukai and Watanabe mailed the third questionnaire to all banks that had been listed on the Tokyo and Osaka stock exchanges in August 1997. The third questionnaire was divided into four categories of accounting, narrowly-defined information, international, and other systems, because the information system specialists at several nationwide banks advised that this categorization was similar to that of questionnaires issued annually by the Center for Financial Industry Information Systems, Tokyo, and was better than the former questionnaires. The reality, however, was that unnecessary complexity was imposed on the responding officers of the information system department of the banks, which had been helpful for the former questionnaires. This situation was contrary to their expectations. The officers of the information system department were required to make their answers coincide with the bookkeeping section, and therefore, the number of responses decreased to 22 banks. This meant that interviews became increasingly important in order to fill the gap between the second and third questionnaires.

At around the time of the third questionnaire (slightly before and after), the wave of bankruptcy engulfed Sanyo Securities Co. Ltd., Hokkaido Takushoku Bank, and Yamaich Securities Co. Ltd. in succession, and the academic curiosity of scientists increased rapidly. This enabled the authors to assemble large-scale workshop, with more than ten members, including economists, accountants, statisticians, and computer scientists, over Tokyo, Osaka, and Kyoto.

In January 1998, Ukai interviewed some information system officers, an acting director, and an assistant director at one of the nationwide banks. As a consequence, it became clear that Japanese nationwide banks were divided in two management groups: centralized information system management and decentralized information system management. In the former group, one particular director of an information system department decided the course of large-scale investment, but in the latter group, a head of information system strategies committee, a managing director or a senior managing director, or a board member had regulatory power concerning the information system and directors of several departments related to the information system determined the department investment by themselves.

This interview added legitimacy to Ukai and his compilation of data concerning management organization and management behavior for information systems since the first questionnaire of 1995.

In February 1998, Ukai set up the second open forum of WISI at the Hankyu Grand Building, Osaka, and reported his simple regressions to three groups of questionnaire answers to the information system specialists at the Daiwa Bank (Resona Bank from 2003) and the Ministry of Finance. Two academic experts in accounting, Prof. Kenji Shiba at Kansai University and Prof. Kazuyuki Suda at Kobe University, joined this forum and provided several helpful comments concerning in consistency between questionnaire answers and financial statements.

In February and March 1998, Hiroaki Aoki, a member of the workshop, interviewed the information system officers at three regional banks in the Kyushu area, based on the discussions of the second WISI forum. He interviewed the acting director of the information system department at one bank, the acting chief of systems development unit at a second bank, and the acting director of the information system department at the third bank. They pointed out that some regional banks had established a joint system development center with other banks, and, therefore, estimation of the value of investment of each bank in information systems was difficult. After this interview, however, the authors discovered that the accounting section at each bank had a better knowledge of the shared costs associated with information system development.

In March 1998, Ukai and Watanabe interviewed the acting director of the information system department and the acting chief of systems development unit at a nationwide bank different to that used in the interview in January. These bank officers clearly mentioned that their bank had invested a large amount of money, being almost comparable to the third-generation on-line system mentioned in Chap. 1, in order to establish continuous ATM services. On the other hand, this bank and another nationwide bank merged into one of the largest Japanese banks and later stopped the 7 days/24-h ATM services in the old former bank branches. Mergers of large Japanese banks do not always improve the efficiency of the information system in the short term, and, on the contrary, can sometimes lead to lower efficiency. This became clear after extensive system problems emerged at Mizuho Bank in 2002. Moreover, the authors were deeply shocked by examples of backward effect that occurred with system integration which saw vast amounts of information system investment from the past left wasted. This practice occurred in other mergers of nationwide banks, and became known as the system integration syndrome in Japanese bank mergers.

The authors estimated simple production functions and cost functions based on the three questionnaires that were sent to the banks and organized the third WISI open forum at the Tokyo Station Hotel in June 1998 to report the estimates. The authors also invited Prof. Iwamura of Waseda University and a researcher at the Center for Financial Industry Information Systems to accept mathematical advice concerning the theoretical background of cost functions of information systems.

In June 1998, Ukai organized the forth WISI open forum at Kansai University Main Library. Koichi Takeda, an associate professor at Hosei University, Tokyo, gave a presentation that described estimates of the production functions of regional banks. It became apparent that the authors' production functions, with three factors, were statistically less robust than that of Takeda with the two factors of capital and labor in a classical approach. The less robust functions were attributed to confusion about the definition of products in the banking industry.

In September 1998, Ukai organized the fifth WISI open forum at Kansai University Main Library. Prof. Kenji Shiba, Kansai University, made a presentation that discussed integration of the information system into the financial system of the bank. The research concerning disclosure of information systems assets started thereafter, being inspired by his Shiba's report, and bore the fruit of Chap. 6.

In September 1998, Ukai organized the sixth WISI open forum at Kansai University Main Library. Hisao Nagaoka, Fellow of the Japan Society for the Study of Office Automation, made a presentation that summarized the history of information systems in the Japanese banking industry. Nagaoka became an associate fellow at the Research Center of Socionetwork Strategies, Kansai University, in 2002.

In addition to these pure academic presentations and interdisciplinary open forums, in April 1999, Ukai interviewed staff from the Research and Statistics Department of the Bank of Japan. From these interviews, it was clear that the Bank of Japan rarely performed statistical analysis of the information systems of individual banks and concentrated on analysis of the macro economy.

The 5 years of research described above came to public recognition, and, in April 1999, a Grant-in-Aid of Scientific Research was awarded to Prof. Ukai as the chief investigator of a project entitled "Development of an Economic Evaluation Method on Information System Investment During Monetary Crises." The members of the project team were Kenji Shiba, Yoichi Iwasa, and Shuzo Yajima, Kansai University; Mitsuru Iwamura, Waseda University; Masayuki Amano, Chiba University; Kazuyuki Suda, Kobe University; Hideki Nishimoto, Ryukoku University; Shinji Watanabe, Osaka Prefectural University; and Koichi Takeda, Hosei University.

As a result of the above-mentioned grant, the Information System Investment Laboratory opened at Tenroku Campus of Kansai University, and an exclusive server was connected to five client personal computers. The server displayed the URL<http://wisi.kutc.kansai-u.ac.jp> and began to disclose the accumulated research of WISI.[5] In addition, editorial meetings for this book were conducted in Tokyo and Osaka, where chapter construction and essential points for the authors were determined.

In March 2000, Ukai visited the office of Prof. Dale W. Jorgenson, Director of Programs on Technology and Economic Policy, J. F. Kennedy School of Government, Harvard University, and President of the American Economic Association at that time, because he sought recognition for originality in the academic society in the USA and Europe. After Ukai's presentation, Jorgenson suggested that information technology analysis in banking industry was original but that Erik Brynjolfsson, Associate Professor at Massachusetts Institute of Technology, had been conducting the same series of analyses as the Ukai group since 1995 into the manufacturing industry. In the following week, Ukai revisited the office of Dr. Daniel E. Sichel, Senior Economist, Board of Governors of the Federal Reserve System, Washington DC, who provided comments similar to those of Jorgenson. After Ukai returned to Osaka, the authors closely investigated Erik Brynjolfsson's research from 1993 to 2000 and concluded that the type of economic analysis that Brynjolfsson used for public data could not be applied to the Japanese banking industry.

At the end of March 2000, WISI mailed a fourth IT investment questionnaire to all banks listed on the stock exchanges in Japan. On this occasion, only six banks returned their answers and the research project was suddenly deadlocked. This was because almost all chiefs of information system depart-

---

[5]URL <http://www.rcss.kansai-u.ac.jp/WISI/>

ments could not afford to reply to the questionnaires after the turmoil of the mergers of Dai-Ichi Kangyo Bank, Fuji Bank, and Industrial Bank of Japan in August 1999, and later of Sumitomo Bank and Sakura Bank, which pushed to the first storm of reconstruction in the Japanese banking industry for 50 years. Therefore, the authors concentrated their research time on statistically analyzing the data of individual banks that had been collected in the first three questionnaires.

In June 2000, Ukai visited the office of Erik Brynjolfsson, then Associate Professor at Sloan School of Management, Massachusetts Institute of Technology (professor from 2001) to discuss the WISI research. Brynjolfsson's main comment was on the reason why Ukai had not taken disclosed data of individual firms in Japan. Ukai was shaken by his misperception of Japanese micro data. Brynjolfsson also gave a pessimistic perspective on IT investment analysis for financial institutions because the effect of IT investment on production was estimated to be smaller in the banking industry than in manufacturing and service industries. He stressed that the IT investment effect was smaller than other production factors, and therefore, this estimation required considerable labor. However, he gave Ukai a copy of Prasad and Harker (1997): a questionnaire survey from the Wharton School, the University of Pennsylvania. This was the biggest benefit of this visit.

Afterward, the authors estimated several production functions using the existing answers to four questionnaires for 1 year. However, all of the statistics in these estimates were very poor. Therefore, a new analytical technique to determine a firm's value through the stock market was used, based on the work of Brynjolfsson and Yang (1997). On the other hand, after the Bank of Japan performed a capital infusion of 7500 billion yen in total, in 15 leading banks in March 1999, the management reconstruction plans for these banks were reported to Financial Supervisory Agency, Japan, and these disclosed some information relating to computer hardware and software. Quite a few estimates of the production function were statistically significant by this disclosure, despite the ambiguous definition of computer hardware and software. This result was described in Ukai and Kitano (2002).

The authors concluded that their accumulated research experience over the past 7 years was applicable to estimates in communication, computer software, and the biotechnology industry. Then a Grant-in-Aid for Scientific Research was awarded to Prof. Ukai as the chief investigator of the project, "Micro-data Analysis of Information System Investment in IT-Related Industries," in April 2001. The members of this project were Motoshige Itoh, University of Tokyo; Akira Sadahiro, Waseda University; Kazuyuki Suda, Waseda University; Shinji Watanabe, Osaka Prefecture University; Hideki Nishimoto, Ryukoku University; and Koichi Takeda, Hosei University. In addition, the associate members were Manabu Shimazawa, Akita University; Fumiko Takeda, Yokohama City University; and Hiroyuki Ebara, Kansai University.

In July 2001, Prof. Akihiko Shinozaki, Kyushu University, reported his research on IT investment at the industry level based on a comparative study of input and output analysis between the USA and Japan. He insisted that development of human resources was the most important factor for IT investment. His conclusion clearly shared common awareness of the issue with the authors.

In September 2001, Ukai and Takemura gave a presentation (Ukai and Takemura 2001) at the seventh WISI open workshop at Shin-Maru Conference Square in Tokyo. In October 2001, Ukai and Watanabe gave a presentation (Ukai and Watanabe 2001) at the Fall Meeting of the Japanese Economic Association at Hitotsubashi University. In each case, the discussion focused on the reason why they had not taken the production function approach.

The Ministry of Education, Culture, Sports Science, and Technology, Japan, had already started the Academic Frontier Promotion Project for Private Universities in April 1996. WISI began a project of comprehensive policy studies into the social effects of information technology in April 2002 under a grant from this program. New participants to the workshop were two computer scientists: Prof. Suguru Yamaguchi, Nara Institute of Science and Technology, Prof. Shinji Shimojo, Cyber Media Center at Osaka University, and an architect, Prof. Akiko Watanabe, Keio University. The Research Center of Socionetwork Strategies was then built in April 2003 and was equipped with the most advanced information technology facilities.

5.2 The Concept of Information System Investment

From the questionnaires, the authors were able to study information system investment in all systems including strategy development, communication network with the outsiders, and in office processing. Thus, information system investment was broadly categorized as (1) expenses for mainframes, workstations, personal computers, and CD machines, and costs of peripheral equipments and/or rental for terminal devices such as CD machines and ATMs, (2) expenses for purchased computer software and charges for computer software, and (3) personnel expenses relating to (1) and (2). This was because other researchers conducted their analyses by dividing the two production factors of capital and labor into the four categories of computer-related capital, non-computer-related capital, computer-related labor, and non-computer-related labor. Moreover, the authors were also familiar with the human capital approach of the Chicago school.

However, from a practical standpoint, there are several different understandings of information system investment in the academic world. First, information system investment was defined in the narrowest way as computer hardware investment. Second, it was defined in a broader fashion as investment in computer hardware, peripheral equipment, and computer software. Third, it was defined in the broadest sense as investment in computer hardware, and computer software, and related personnel costs. The authors of this book shared a common awareness of the issue in taking the widest definition of investment including human resources. In the other words, it was the total value including computer hardware, peripheral equipment, all of the communication network including ATMs and terminals, computer software, and personnel cost of development and maintenance. In this book it is described as "Information System Investment I." The stock concept of this investment is described as "Information System Asset I."

The alternative treatment of analyzing the total value of computer hardware, peripheral equipment, all communication networks of ATMs and terminals, and computer software was also considered. In this book, such an approach is described as "Information System Investment II." The stock concept of this treatment is described as "Information System Asset II."

The third questionnaire divided banking information system into four categories of accounting, narrowly-defined information, international, and the other system. This narrowly-defined information system includes strategic planning, product development, and customer services with computer network. The total value of investment for this kind of information system is described as "Information System Investment III." The stock concept of this investment is described as "Information System Asset III."

Finally, "computer hardware investment" was defined as the total value of the computer hardware and all of the communication networks of ATMs and terminals. Similarly, "computer software investment" was also defined as the total value of computer software investment.

On the other hand, the business persons who answered the questionnaires took serious issue with these definitions of information system investment as discussed above. The details of some of their comments are given below as useful suggestions for the research of economists:

1. An information system officer commented as follows. "Operation of the system includes system function upgrades, computer operation, reducing ledger sheet sizes, office procedures, decisions related to operating procedure, and manual tasks. Generally speaking, system renewal may consist of total renewal, partial renewal, or particular functional upgrades by extension of production line. Therefore, it is difficult to formally answer the questionnaire."

2. An information system officer at a second bank commented as follows. "The bank has not renewed its information system as regularly as other banks, for example, in the second-generation or third-generation on-line system renewals. Therefore, I cannot answer the questionnaire under your definitions and almost all questions are not applicable."

3. An information system officer at a third bank commented as follows. "The existing information system consists of several thousand programs and these have development periods and operation times that are quite different. Moreover, we always conducted maintenance including function additions, improvements, and upgrades after the operation started. There-

fore, I cannot answer you precisely. Your questionnaire's concept would be ambiguous for information systems in the banking industry."

4. An information system officer at a fourth bank commented as follows. "In spite of your interesting questionnaires, our bank has been operating only the accounting system and a partial information system within the accounting system and has not constructed an independent information system. However, we formed an information system development task force last year and are now conducting fundamental examinations to begin operation of the information system in 1997, because the system will become more important as Japanese financial deregulation proceeds. We have been codeveloping our computer software with another bank since 1981."

(It was later established through a telephone interview that this bank dismissed their task force in April 1996, because of difficult economic circumstances.)

5. An information system officer at a fifth bank commented as follows. "Our bank operates the information system by outsourcing to our related company; therefore, we are unable to answer the questionnaires."

(This type of outsourcing is rapidly increasing among regional banks.)

6. An information system officer at a sixth bank commented as follows. "I answered the questionnaire concerning the most recent individual information system except for question 5-1. For question 5-1, I summed the rental and lease charges and maintenance costs of the entire system in our bank."

7. Some questions concerning employment conditions and personnel input may have been confusing for the respondents. An information system officer at a seventh bank commented as follows. "I answered in terms of a cumulative total number of employees in man-months, because the number of information system personnel always changes."

(This type of man-month measurement of labor input was used by the authors after the second questionnaire.)

5.3 Accounting Problems in Information System Investment

After careful examination of respondents' comments to the first questionnaire, the authors conducted supplementary interviews with IT section officers. In this process, the authors were at a loss to decide who or what section was in charge of entire information systems in Japanese banks. In particular, the authors were shocked to find that they could not identify who was the Chief Information Officer (CIO), the top specialist for information systems, in several Japanese banks. In addition, section officers indicated that an expenditure to be transferred to investment, on accounting principle, was transferred to current expenses in reality. This procedure cannot be described by financial statements.

Based on this experience, the authors provocatively asked the respondents whether they might be confused over investment cost and current expenses in the business field, in the last part of the second questionnaire. The respondents' reactions to this question were very interesting, and some of these are listed as follows.

1. A bank officer made the following comment. "The orders to outside manufacturers and software engineers and outsourcing costs are processed exactly. On the other hand, some of our inside employees are filling dual roles in the development section and the operation section. This kind of cost could not be clearly divided. However, we have not answered this type of questionnaire precisely."

2. A second bank officer commented as follows. "At this stage, we expect no great innovation in the information system, except the introduction of all-day-and-night operation of banking business. Therefore, the small development cost might be transferred into operation costs. The difference of average age does not reflect the difference of annual income, because over time the wages of the younger generation makes up for the pay gap with the older generation."

3. A third bank officer commented as follows. "I understand your provocative question. I also had trouble differentiating between the investment cost and current expenses for the purpose of answering your questions. I defined operating section employees as operators, punchers, and system controllers, who are responsible for filing, database compiling, and bookkeeping, and classified development section employees as persons in charge of infrastructure of the on-line system. However, even in the operating section employees are conducting development and upgrading operation tools, schedule controllers, and job control cards. Therefore, the division between the development and operational sections will become more difficult as information system techniques progress. Similarly, the development of the personal computer makes it more difficult to distinguish system engineers from ordinary clerks."

4. A fourth bank officer commented as follows. "Your trouble will derive from the understanding of cost. We can neither control nor understand cost definition clearly when a bill statement is received that mixes the development and operation of the information system. Suppose you outsource the entry of test data and ordinary running data together; you will not be able to clearly divide the electricity charge into the development cost and the operation cost. It will be the same when using the same computers for system development and system operation."

5. A fifth bank officer commented as follows. "In the system development section, the machines are shared with the ordinary operating section; therefore, it will be difficult to understand the different costs without

a cost control system. Personnel costs and machine usage costs will vary widely changed according to whether system operation costs include system maintenance costs in the development section. In this regard, it likely that you will not be able to clearly understand the costs. In our bank's answer, development costs include only the development cost for the new information system, and the operation costs include personnel cost of the operation section and the outsourcing cost to the joint system operation center. Although this outsourcing cost might be a sum of development and operation costs, our bank regards it as operation cost totally."

6. A sixth bank officer commented as follows. "Our bank allocates the employees depending on service and production classification. Therefore, development costs including addition of new functions and new production development will not be able to be differentiated from operation costs including code system changes and file size enlargement. This kind of rigid cost control is not always necessary."

7. A seventh bank officer commented as follows. "I answer you on the understanding that the information system operation includes system maintenance and ordinary system development. Most of our system engineers are in their late twenties, and were employed after 1985. Almost all of them came from nonscientific colleges and had accepted half-year training positions in computer science."

8. An eighth bank officer commented as follows. "It depends on the respondent's background to understand the difference between information system development costs and operation costs. If these definitions were clearer, the answers could be more complete."

9. A ninth bank officer commented as follows. "It is difficult to understand the information system costs in terms of identifying development and operation costs. In other public questionnaires, almost all classifications of information system costs were the rental rates of host computers including computer software, the equipment costs of computer centers including depreciation of movable and immovable assets, the depreciation of equipment in banking offices, the cost of supplies, networking costs, and personnel costs in the system development section and the operation section. This type of classification makes it difficult to divide the above-mentioned costs into development and operation costs."

However, on proceeding with interviews with system department officers, the authors realized that these kinds of ambiguous answers were mainly returned by small regional banks. It was also found that system officers at some nationwide mega banks definitely stated that development and operation costs were divided and controlled clearly.

Moreover, the answer of one nationwide mega bank was that its system development costs were classified and calculated as new computer hardware and software costs and its operation costs were classified into several running costs including repair and maintenance costs.

Another nationwide mega bank provided the following remarks. "Because the outsourcing cost of information system development and the cost for purchasing machines and equipment are transferred to the asset item from the standpoint of business accounting, development costs should definitely be classified apart from system operating costs. Even in the case of information system developed exclusively by insiders, the personnel costs in the development and operation sections should be clearly distinguishable."

A third nationwide mega bank provided the following remarks. "It is essentially impossible to mix system development with system operation. These two classifications should be clarified. First, a total information system has to be clarified with the definition and classification of development and operation. Second, administrative setup and job demarcation must be clear and logical."

As a result of the comments of a number of mega banks, the authors concluded that the accounting confusion between system development and operation was very rare, at least in Japanese nationwide mega banks.

5.4 Lessons from Questionnaire Surveys

To allow the effect of information system investment on productivity to be statistically analyzed by micro data, it is necessary to put in place a comprehensive database of computer hardware, peripheral equipment, computer software, knowledge asset concerning system development, and related personnel costs in individual firms concerning information system investment. Economist will interpret knowledge and human resource as some kind of capital. On the contrary, accountant will never do it. This book takes the economist approach throughout Part III.

The Japanese situation concerning computer software has certainly improved since 1999, as is described in detail in Chap. 6. However, these computer software data have never been disclosed with the hardware and related personnel costs, because almost all executive officers in each firm have no interest in comprehensive IT data.

On the other hand, in the modern democratic society, the extent of information that firms now disclose is increasing even without the obligation of accounting rules due to increasing social pressures. In today's economy, information technology is extremely important, therefore, it is alarming that information technology investment is not disclosed at the individual firm level, considering that an exact measurement of national power generates proper national policies.

Speaking more academically, it would be a shameful situation for a highly industrialized nation like Japan that government agencies and big enterprises import information technology investment data from research institutes in the US and Europe. Japan should be capable of making major contributions to international information technology policy.

The authors suggest that every company listed on the Tokyo, Osaka, and Nagoya stock exchanges should disclose their details of information technology investment. Two or three neutral research institutes in the private sector should disclose their rankings of disclosure in information technology investment of listed companies.

Moreover, even though financial statements are likely to disclose adequate information of information system in the near future, it will be necessary for social scientists to continue to issue the kinds of questionnaires used in this study. The 8-year survey by the authors has taught them that mathematical economists and econometricians should always pay attention to factors that are not disclosed in financial statements.

6

Disclosure and Circumstances Concerning Information System Assets

T. Takemura, Y. Ukai, and H. Nagaoka

6.1 Introduction

Cooperate information disclosure in Japan had not caught up with European countries and the United States until the later half of 1990's. In the progress of investors' relations and economic globalization, it was urged to establish the information disclosure system within Japan. Creating some legal regulations and establishing disclosure systems have been progressing over the past few years. On the contrary, the definition s and range of information system assets are too vague, and the cooperate executives do not recognize the importance of information system assets. This situation was a bottleneck of the economic analysis of information system investment at firm-level. However, in the later half of 1990's the accounting standards for computer software established and the authors could examine their economic effect.

The Business Accounting Council (Ministry of Finance) implemented the "statement of opinions on the establishment of accounting standards for research and development costs" in March 1998. In the statement of opinions, the accounting standards for research and development costs (the authors call it "1998 R&D standards" hereafter) clearly defines computer software for the first time.[1] Despite these standards for computer software, many banks in Japan still do not mention computer software in their financial statements. Thus, the authors outline those banks disclosing their information systems, and review the circumstances in the Japanese banking industry.

This chapter consists of the following sections. Sect. 6.2 sketches accounting procedures for information system assets in the financial statements. Sect. 6.3 explains the reason why computer software is regarded as asset. Sect. 6.4

---

[1]The Financial Services Agency has been supervising the Business Accounting Council since September, 2004.

briefly outlined the disclosure systems. Sect. 6.5 describes the circumstance concerning information system assets in the Japanese banking industry. Sect. 6.6 classifies banks by whether or not they disclose information system assets in their financial statements. Finally, Sect. 6.7 presents themes for further research.

6.2 Accounting Procedure of Information System Assets

6.2.1 Accounting Procedure (1): Computer Software Assets

The Business Accounting Council (Ministry of Finance) implemented the statement of opinions on the establishment of accounting standards for research and development costs in March 1998. After receiving the statement of opinions, the council revised the accounting rule for terms and presentation of financial statements in November 1998. Furthermore, the Japanese Institute of Certified Public Accountants implemented the "practical guidelines on accounting standards for research and development costs and software production costs" (the authors call it "practical guidelines on computer software" hereafter) in March 1999. They were applied from April 1999. Until the accounting standards and the practical guidelines were implemented and applied, no clear definition of computer software existed except for Corporation tax law, primary regulation notice 8-1-7. Therefore, each firm applied its accounting procedure for computer software arbitrary.[2] The 1998 R&D standards prescribes as following. If software is applicable to research and development, it should be charged as cost (current expenses) and should be written as research and development cost (current expenses) in the financial statements. Otherwise, computer software is included intangible assets and is written as computer software asset in the financial statements by using a procedure for depreciation on computer software prescribed in the practical guideline.

With the development of information and communication technology (ICT), the accounting standards have been changed in the progress of information age: the importance of computer software, and comparability of financial statements. Sect. 6.4 reviews the characteristics of computer software assets and the international comparability of financial statements. Chap. 9 will statistically analyze the computer software.

---

[2]Corporation tax law, primary regulation notice 8-1-7, presented the following notice until April 2000: the computer software that firms purchased from the outside and/or entrusted companies having an effect of more than one year should be treated as deferred assets and should be depreciated in 5 years when its value is more than 200000 yen. However, the notice did not regulate development costs of computer software made by themselves or for sale on the market. In addition, by a current technological advance and changes in management environments, various problems were pointed out by people engaged in business. This notification was abolished by computer software taxation business-related revision of the tax system.

6.2.2 Accounting Procedures (2): Computer Equipment

Information system assets except computer software (called "computer equipment" in this chapter) include various computers and instruments such as mainframe computers, office machineries, personal computers, and peripheral equipment, instruments related to cash dispenser (CD) or automatic teller machines (ATMs), and terminal equipment for on-line systems.

Since most computer equipments took physical forms, they are written as tangible fixed assets in the financial statements.[3] When the computer equipment is in the financial statements as tangible fixed assets, they should be depreciated along their useful life. Generally speaking, depreciation period is 6 years for computers, 10 years for communication instrument devices, and 5 years for personal computers.[4] However, Moore's Law[5] (regarding integrated circuit technology) and "dog year[6]" (concerning the Internet) represent the high speed of progress in information technology. Therefore, it is disputable that these accounting procedures reflect the reality. Of course, in the case of functional obsolesce by technological progress, such computer equipments are able to be depreciated in acceleration.

In the early 1990s, each firm did not only hold computer equipments as asset, but also took advantage of lease and rental contracts. It was because lease and/or rental contracts are efficient and convenient for the management stability.[7] Indeed, many banks tended to lease computer equipments in late 1990s.

In June 1993, the First Subcommittee of Business Accounting Council (Ministry of Finance) implemented the "statement of opinions on accounting standards for lease transactions." Until its implementation in April 1994, leasing computer equipments was not included in the balance sheet of borrowers.[8] In other words, lease transactions were considered an off-balance transaction,

---

[3]Tangible fixed assets are assets that each firm owns for more than one year for the purpose of use, has physical form, and its value exceeds a fixed sum. Intangible fixed assets do not have physical form, are used in management for a long term, and provide some economic profit.

[4]See Ito and Kotani (1999).

[5]Moore's Law is a law concerning integrated circuit technology and is named after Gordon Moore, who was a cofounder of Intel Corporation. Essentially, the law states that the calculation velocity of an integrated circuit doubles every 18 months.

[6]Dog year is an expression that is used to describe the phenomenon in which improvement of computers in the present age occurs earlier than before and seem to advance by around 7 times as sense.

[7]Leases and rentals imply hiring machines or equipments. The former are transactions in which the borrower chooses a starting date and the leasing company supplies the machine which the borrower appointed beforehand. The latter is a transaction in which borrowers are an unspecified number of the general public and lender loans the same machine to various users in several times.

[8]"Practical guidelines on accounting standards for lease transaction and the disclosure" was announced by the Japanese Institute of Certified Public Accountants.

and computer equipments were not included in the assets on the financial statements. After the implementation of the standards rental and/or lease of computer came to be explained on the notes to the financial statements.

The lease transactions are classified into "financial lease transactions" and "operational lease transactions." The former satisfies the following three conditions: (1) the lease contract is not able to cancel during the lease period, (2) the borrower can receive his or her benefit from leased objects, and (3) the borrower substantially covers cost related to use the lease objects. The transaction that does not satisfy the above conditions is called operational lease transaction.

6.2.3 Accounting Procedure of Research and Development Costs

Research and development is the driving force of firm's growth, and is one of the important factors for management strategy and/or the future profitability in any business. Generally speaking, research and development are mentioned on "firm's general conditions" in the financial statements, but each firm arbitrarily applied it to accounting procedures as well as computer software.

The statement of opinions on the establishment of accounting standards for research and development costs clarifies their definitions, and accounting procedure for research and development. The statement on opinions mentions that expenditure on research and development was not appropriate to be mentioned in the balance sheet and should be treated as cost (current expense). A reason is as follows. Whether they will lead on to profit in the future is unforeseeable, and even if a research and development plan advances, and expectations for future profit grow, the realization of the future profit is still not assured.[9]

Note that software production regarded as research and development, even if accounting procedures are applied to cost (current expense).

6.3 Computer Software as an Asset

As shown in the previous sections, arbitrary accounting procedures (expense or asset) on computer software had been applied until the 1998 R&D standards. In this section, the authors mention computer software from the standpoint of asset. From this point of view, the authors will use computer software asset for analysis in Chap. 9.

In March 1992, the Software Information Center (SOFTIC) announced the "computer software accounting practical guidelines," called SOFTIC plan. The SOFTIC plan asserted that computer software essentially had a nature of asset in the future profitability. Furthermore, the SOFTIC plan requires the

---

[9]See Appendix B in details.

following criteria on the nature of asset: "technological feasibility" and "market feasibility" to computer software for sale on the market, and "technological feasibility and usefulness" to compute software for internal use.[10]

In October 2000, the System of National Accounts (SNA) was revised in Japan from 68 SNA to 93 SNA. In this revision, 93 SNA defined that the gross fixed capital formation includes custom-made computer software which firm purchases.[11] K Matsumoto (2001) provides a summary of USA-Japan comparison of IT-related statistics at an aggregate level.

The SOFTIC plan, SFAS No.86, and 93SNA clearly allow the nature of asset concerning computer software.

6.4 Types of Cooperate Disclosure

6.4.1 Legal and Voluntary Disclosure

Disclosure systems are classified into two types; legal disclosure and voluntary disclosure.[12]

Legal disclosure is implemented by Law, such as securities and exchange law and/or commercial code. Each law requires that listed firms should continuously disclose their information including business performance and financial status.[13] For example, one of legal disclosure is submission of financial statements.

Furthermore, securities and exchange law and/or commercial code were revised for several times, and the legal disclosure system was improved. The Agency of Securities Commissions have operated disclosure system on the

---

[10]SOFTIC plan is influenced by USA Statement of Financial Accounting Standards No.86 (SFAS No.86), "the Accounting for the costs of computer software to be soled, leased, or otherwise marketed", which is useful for judging the asset characteristics of computer software for sale. In this issue, computer software has asset character if it satisfies the following conditions: (1) market feasibility, (2) financial feasibility, (3) technological feasibility, and (4) management commitment. See Ito and Kotani (1999).

[11]In 68 SNA, the definition of computer software, which becomes inseparable from the computer hardware, was included in total fixed capital formation because the estimation of computer software is impossible separated from the computer hardware. The other computer software (for example, computer software which firms accept orders) was treated as intermediate consumer goods, and it was not included in gross domestic product.

[12]The effectiveness of disclosure system is determined by four factors; (1) quality of information, (2) quantity of information and the scope of information, (3) timing of disclosure, and (4) procedure of disclosure. These factors are interdependent with each other.

[13]Securities and exchange law have the fundamental principles of protecting investor from disbenefit, and commercial code has the fundamental principles of protecting creditors.

Internet, called EDINET (Electronic Disclosure for Investors' Network) since June 2001. This electronic disclosure of financial statements was obliged from April 2004.[14]

Voluntarily disclosure is implemented by each firm. In recent years, each firm begins to disclose information if they judge it to be useful. They can announce their information in pamphlets and/or by Internet, and write it in their financial statements.

For example, each firm submits the financial statements by legal disclosure, but it is not enough to analyze the information technology system. Then they can voluntarily disclose the IT information. That is, they can voluntarily write it on "situation of equipment," "financial situation," or "firm's general conditions" in their financial statements, if they regard it as an important factor. In this chapter and Chap. 9, the authors focus on the banks which disclose the information system.

Such voluntary disclosure is called investor relations (IR) and plays more important role in recent years. From early 1990s interest to IR was still high in firms inside and outside. Thus, in Japan, the Japan Investor Relations Association (JIRA) was established in May 1993 after the model of the National Investor Relations Institute (NIRI) in the USA.

Differences between legal disclosure and voluntary disclosure are as follows. At first, the former requires firms to disclose their information only within the framework of the regulations, while the latter has no obligation. In other words, voluntary disclosure can provide immediately information meeting the user needs. Second, legal disclosure can bring a charge against the firm violating laws or making misstatements. On the other hand, the voluntary disclosure system can impose no such penalties. If the information disclosed is far from the reality, firm concerned would suffer in the market. For example, the firm may be excluded from the stock market.

6.4.2 Necessity of Disclosure

Firms can make reliable relations among other firms by providing information regarding business performance, description of business, and the policy. From viewpoint of legal disclosure, firms listed on stock markets are obliged to disclose true information not only for stockholders and investors, but also for the society. Unlisted firms need to disclose their information if they pay attention to potential investors.

Even if firms submit formal financial statements, it is not enough to fulfill obligation of top executives in recent years. In particular, when stockholders and investors analyze or forecast the potentiality and management strategy,

---

[14]In the USA, the Securities and Exchange Commission (SEC) established the EDGER database in May 1996. Anybody can easily and instantaneously have access to firms' financial information database.

additional information becomes important factors for them. At the same moment, if firms disclose useful information for society, firms can capture potential investors.

However, the authors should remark that disclosure does not have always advantages for the management. Even though it leads to disadvantages, firms concerned should unveil the information for market economy.

Disclosure system is needed for international comparability of financial statements. In 1987, the International Accounting Standards Committee (IASC: the International Accounting Standards Board, IASB since 2001) started a project called comparability of financial statements. The aim of this project is controlling arbitrary accounting procedure accepted and unifying accounting standards in financial statements. For example, if each firm makes financial statements in agreement with arbitrary accounting procedure, international and domestic comparability of financial statements would be consequently impossible and it is difficult to do comparative analysis. The information technology is a typical case for this issue.

When the authors solve various problems or investors analyze firms, information concerned is required. Therefore, the information has to be written in financial statements. Especially, it is expected to disclose information concerning research and development, and information system on their financial statements. Information system is important strategy of management, as the authors described in Chap. 2.

6.5 Information System Assets in Japanese Banks

In this section, the authors concretely view actual state concerning information system in Japanese banking industry. The data is constructed from financial statements each banks submit in the 1993-1999 period. Almost banks are listed on the first section and the second stock exchanges (in Tokyo, Osaka or Nagoya), or the local stock exchanges. (in Sapporo, Niigata, Kyoto, Hiroshima or Fukuoka)[15]

Even though some banks newly list on the market, fail, or merge with other banks, the number of banks fluctuated by approximate 4%. Table 6.1 gives the number of banks in Japan.[16] Matsuura, Takezawa and Toi (2001) provide a summary of failures or mergers of financial institutions in Japan.

The authors explain our data set, and the statistic in the following. First of all, in the 1993-1999 period, the authors examined whether data concerning information system assets were written in financial statements.[17] Computer

---

[15]Both the Niigata and Hiroshima stock exchanges were merged in March 2000 with the Tokyo stock exchange, and the Kyoto stock exchange was merged with the Osaka stock exchange in March 2001.

[16]Note that over-the-counter banks and foreign firms listed on the market are not included in our data set.

[17]See Takemura (2002).

Table 6.1. Number of Japanese banks in the 1993-1999 period

	1993	1994	1995	1996	1997	1998	1999
Number of banks	108	109	111	116	114	114	112

equipment is written on "situation of equipment" in financial statements; most banks hold a part of computer equipment such as office machineries as tangible fixed assets, and they leased computer equipment such as personal computer, general-purpose computers and on-line system terminal devices. On the contrary, few banks wrote computer software in financial statements although they use computer software.[18] Even though the number of sample decreases, authors can describe the circumstances concerning information system assets in Japanese banking industry.

Fig. 6.1 shows the changes in the number of banks in Japan based on charging computer software in the 1993-1999 period: (1) banks write computer software as an asset in the financial statements, (2) banks write computer software as cost (current expense) in the financial statements, and (3) banks do not write computer software in the financial statements.

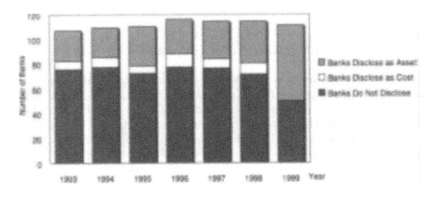

Fig. 6.1. Changes in the number of banks in Japan based on charging computer software

Before the 1998 R&D standards was implemented, the proportion of the banks posting information about computer software on "financial situation"

---

[18]It is possible that computer software are written on "firm's general conditions" in financial statements includes R&D, but no banks where wrote the general condition of R&D in the 1993-1999 period. Actually, the authors cannot imagine that each bank does not conduct R&D.

in financial statements was only average 33% in 1993-1998 period. Among banks charging computer software, average 78% of the banks charged as an asset. The remaining banks charged them as cost (current expense) in the 1993-1998 period. On the other hand, in 1999 (after the 1998 R&D standards was implemented) the proportion of banks charging computer software rose to around 55%, and all the banks charged computer software as an asset. Implementing 1998 R&D standards and management's accounting decisions imply that the management executive of each bank knows the importance of computer software as an asset.

Next, Fig. 6.2 shows the proportion of banks based on the frequency of writing computer software in the financial statements more than one times in the 1993-1999 period.

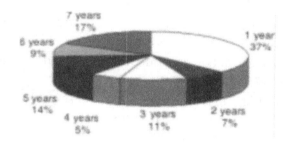

Fig. 6.2. Proportion of banks based on the frequency of writing computer software in the financial statements more than one times in the 1993-1999 period

Although 35 banks (the proportion is about 30.2%) did not write computer software in the financial statements, 81 banks (the proportion is about 69.8%) wrote computer software in the financial statements more than one times. In Chap. 9, authors will construct panel data set with 74 of 81 banks charging computer software as an asset. In 1999, 24 banks (the proportion is about 20.7%) wrote computer software as an asset in the financial statements for the first time. Although the accounting procedure of computer software was clarified by the 1998 R&D standards, some banks would still choose not to write computer software in the financial statements. Authors will discussed the banks choosing not to write computer software in the financial statements further in the next section.

6.5.1 Information System Assets

Table 6.2 shows summary statistic of the information system assets for the banks in Japan described above.[19]

Table 6.2. Statistics of information system assets in Japanese banks for each year during 1993-1999

Year	Average*	Median*	SD	Minimum*	Maximum*	Number of banks
1993	13026.76	7711	15282.00	961	56490	25
	2.61	2.31	1.36	0.51	5.21	
1994	10651.08	7111.5	12685.49	1143	48975	24
	2.32	2.17	1.06	0.61	4.23	
1995	17086.97	6497	21245.67	448	75294	32
	2.55	2.46	1.26	0.53	6.19	
1996	20059.21	5743.5	32268.91	1058	144991	28
	2.67	2.26	1.69	0.82	9.32	
1997	21101.13	6238	29898.21	1073	88845	30
	2.87	2.56	1.44	0.68	5.75	
1998	22509.18	7207.5	31937.61	1114	97614	34
	3.12	2.67	1.54	1.07	6.34	
1999	17269.64	5534	26735.10	476	102799	61
	3.13	2.43	2.26	0.52	13.94	

Note that the lower in each row represents the system asset value per employee
* one million yen

In Table 6.2, the average amount of information system assets exceeds about 10 million dollars, and the medians are 5.5-7.7 million dollars in the 1993-1999 period. This difference between average and median is caused by a few banks (nationwide banks) holding a large amount of information system assets. In 1993, the proportion of banks holding information system assets valued at 20 million dollars or less was approximate 80% among banks charging computer software. In the 1994-1999 period, the proportions of the banks were 88%, 75%, 75%, 88%, 88%, and 80% in sequence. On the other hand, in the 1993-1999 period, the proportions of banks holding information system

---

[19]In Table 6.2, note that the authors use information system assets treated in accounting procedure as an asset. Refer to Chap. 9 for details of information system assets used in this chapter.

assets valued at 40 million dollars or more were 12%, 8%, 16%, 18%, 20%, 20%, and 15% in sequence.

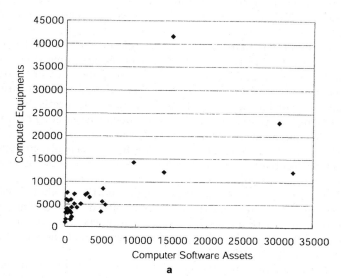

a

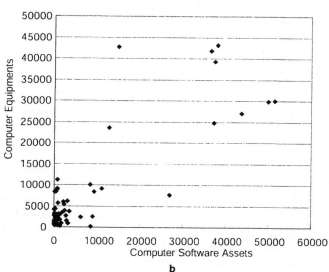

b

a) 1993 and b) 1999

Fig. 6.3. Scatter diagram between computer software assets and computer equipment for Japanese banks

Furthermore, in the 1993-1996 period, the proportion of banks holding information system assets per employee valued 40 thousand dollars or less was average 85%, and the proportion in the 1997-1999 period dropped to approximate 75%. The median of information system assets per employee were around 21-27 thousand dollars and the average tends to rise slightly after 1994.

The authors find that share of information system assets in total assets was kept at constant 0.19% in the 1993-1999 period. At the same time, because the information system assets are composed of computer software assets and computer equipment, the authors finds that share of computer software assets and computer equipment in the total assets are slight, respectively.

In Fig. 6.3 authors plotted relationships between computer software assets and computer equipment in 1993 and 1999. Banks mass on the periphery of point of origin (at the lower left) in each graph. Especially, in 1999, it is notable that several banks hold both computer software assets and computer equipment valued at 10 million dollars or more.

In the 1993-1999 period, many banks (the proportion is about 78%) held computer equipment more than the computer software assets. The share of computer software assets in information system assets was around 33% on average in the 1993-1999 period. Incidentally, averages of the share in 1993 and 1999 were 33.3% and 30.0%, respectively.

The authors can treat information system assets separately divided into computer software asset and computer equipment.[20] Thus, we divide information system assets into computer software assets and computer equipments in the following subsections, and view them.

6.5.2 Computer Software

Table 6.3 shows summary statistic of computer software for the Japanese banks in the 1993-1999 period. Note that we categorize computer software charged as an asset into one charged as cost (the current expense) specifically in this table.

The medians of computer software assets were 1.5-2.1 million dollars in the 1993-1998 period, but it dropped to about 1 million in 1999. We guess that accounting procedure was changed by implementing 1998 R&D standards. That is, banks charging computer software as cost (current expenditure) should change to do as an asset suddenly. In addition, the other banks that previously had not write computer software in the financial statements also should change their policy and decide to charge it.

The average amount of computer software assets had risen year after year in the 1993-1998 period, but dropped by 25% in 1999.The authors find a similar tendency in the case of computer software assets per employee. As

---

[20] As shown in Sects. 6.2 and 6.3, each accounting procedure is different and the computer software has a special property.

Table 6.3. Statistics of computer software in the 1993-1999 period

Year	Category	Average*	Median*	SD	Minimum*	Maximum*	Number of banks
1993	Asset	5715.80	2152.00	8682.18	43	32194	25
		1.05	0.68	0.98	0.02	3.25	
	Cost	317.43	293	232.27	104	806	7
		0.13	0.10	0.14	0.04	0.44	
1994	Asset	4268.13	1907.50	6695.46	105	30649	24
		0.87	0.55	0.76	0.07	2.84	
	Cost	333.63	272	201.83	125	708	8
		0.13	0.12	0.08	0.05	0.27	
1995	Asset	6246.59	1909.50	9011.03	7	31139	32
		0.81	0.57	0.75	0.01	2.90	
	Cost	1493.83	489	2032.36	105	5246	6
		0.41	0.19	0.65	0.05	1.73	
1996	Asset	6790.04	1573.50	10599.28	29	33970	28
		0.85	0.60	0.76	0.02	2.55	
	Cost	2825.55	247	7238.18	20	24513	11
		0.32	0.13	0.47	0.03	1.62	
1997	Asset	7863.20	1713.50	13057.67	30	47324	30
		0.94	0.78	0.78	0.03	2.57	
	Cost	682.13	267	929.25	46	2805	8
		0.19	0.15	0.14	0.07	0.49	
1998	Asset	9690.91	2123.00	15545.74	31	54624	34
		1.18	0.81	0.94	0.02	3.58	
	Cost	705.33	375	927.56	50	2878	9
		0.19	0.17	0.10	0.04	0.34	
1999	Asset	7299.20	999.00	13560.62	0	51455	61
		1.07	0.60	1.25	0.00	5.61	
	Cost	—	—	—	—–	—	—
		—	—	—–	—	—	

Note that the lower in each row represents the system asset value per employee
* one million yen

shown in Table 6.3, we had a large difference between the average and the median in the 1993-1999 period. The proportion of banks holding computer software assets valued at 10 million dollars or more was around 20%, and the proportion of banks holding computer software assets valued at 2 million dollars or less was around 55%.

The average of computer software cost (current expenditure) had risen year after year in the 1993-1996 period, but it dropped in 1997. By implementing the 1998 R&D standards, no banks charged computer software as cost (current expenditure) in 1999.

6.5.3 Computer Equipment

Table 6.4 shows summary statistic of computer equipment for the Japanese banks in the 1993-1999 period.

Table 6.4. Statistics of computer equipment in the 1993-1999 period: I

Year	Average*	Median*	SD	Minimum*	Maximum*	Number of banks
1993	7310.96	5007	8341.25	868	41384	32
	1.56	1.59	0.61	0.49	2.79	
1994	6382.96	4134	6973.43	1016	32543	32
	1.45	1.40	0.59	0.54	3.02	
1995	10840.38	5261	13104.86	441	52510	38
	1.74	1.62	0.99	0.52	5.80	
1996	13269.18	4092	23809.96	905	118835	39
	1.82	1.50	1.33	0.40	7.64	
1997	13237.93	4753	17445.75	894	56830	38
	1.93	1.78	0.89	0.39	3.76	
1998	12818.26	4603	17067.87	1044	59335	43
	1.94	1.81	0.82	0.56	3.69	
1999	9970.44	3696	14364.72	463	65443	61
	2.06	1.79	1.47	0.22	10.17	

Note that the lower in each row represents the system asset value per employee
* one million yen

The computer equipment is applied to accounting procedures as follows: the computer equipment such as office machineries are charged as a tangible asset, and computer equipment leased such as peripheral equipment and personal computers is not charged as an asset. Then, the computer equipment

Table 6.5. Statistics of computer equipment in the 1993-1999 period: II

Year	Category	Average*	Median*	SD	Minimum*	Maximum*	Number of banks
1993	Asset	5836.19	4248.50	6573.63	682	35029	32
		1.37	1.29	0.59	0.36	2.61	
	Cost	1297.97	973.00	1350.91	0	6645	
		0.34	0.36	0.20	0.00	0.74	
1994	Asset	4629.16	3459	4651.62	536	22715	32
		1.18	1.14	0.57	0.28	2.86	
	Cost	1377.84	985	1742.38	0	8704	
		0.37	0.35	0.25	0.00	0.90	
1995	Asset	7982.08	3918.50	10429.55	348	41973	38
		1.31	1.19	0.95	0.23	5.80	
	Cost	1854.97	1034.00	2837.23	0	15639	
		0.38	0.34	0.29	0.00	1.12	
1996	Asset	8927.31	3774.00	17307.10	319	97490	39
		1.24	0.93	1.03	0.25	6.26	
	Cost	1745.03	833.00	2865.06	33	14801	
		0.34	0.29	0.27	0.01	0.89	
1997	Asset	7858.03	3145.50	11818.85	344	43049	38
		1.17	1.00	0.72	0.30	2.95	
	Cost	1644.84	802.50	2914.73	33	15173	
		0.33	0.29	0.25	0.01	0.91	
1998	Asset	7483.56	2676.00	10954.14	329	42427	43
		1.15	1.17	0.64	0.27	3.04	
	Cost	1601.44	779	2515.94	0	12939	
		0.34	0.35	0.23	0.00	0.96	
1999	Asset	6901.84	2296.00	10714.71	0	42685	61
		1.35	1.08	1.41	0.00	10.17	
	Cost	1014.93	427.00	1849.82	0	12753	
		0.28	0.16	0.33	0.00	1.30	

Note that the lower in each row represents the system asset value per employee
* one million yen

leased (on the financial lease transactions) was accumulated by lease of each item after 1993.[21]

The authors find that we had a large difference between the average and the median in the 1993-1999 period, but the difference between them per employee was so little. Furthermore, the authors divide computer equipment into one leased and one charged as an asset, and view them separately.

Table 6.5 shows summary statistic of computer equipment for the Japanese banks in the 1993-1999 period with computer equipment charged as tangible fixed asset and computer equipment on the financial lease transaction.[22]

In almost banks (the proportion is around 90%) the share of tangible fixed asset such as office machineries in computer equipment exceeded 50% in 1993-1994 period. The proportion decreased to around 85% after 1995. It is apparent that banks tended to prefer to lease computer equipment rather than to hold it as an asset.

Note that some minimums of computer equipment in Table 6.5 are zero. This implies that the banks do not write computer equipment on either procedure in the financial statements.

The median of computer equipment on each procedure decreased after 1995, and we have a large difference between the average and the median in the 1995-1999 period. The average and median of computer equipment charged as an asset per employee were approximate 12-13 thousand dollars. At the same time, the average and median of computer equipment on financial lease transaction per employee was about 3.5 thousand dollars.

6.5.4 Planned Budget for Information System

In the financial statements the planned budget for information system is written on "situation of equipment." This budget will be expended for the purposes of planning in the next fiscal year.[23] Most banks disclose the information in the financial statements.

Table 6.6 shows summary statistic of the planned budgets concerning information systems of the Japanese banks in the 1993-1999 period.

In the 1996-1998 period, both the average and the median of the planned budgets concerning information system tended to decrease. On the other hand, the average and median of the planned budgets per employee did not change drastically. However, the median of the planned budgets in 1999 suddenly decreased one thousand dollars from ones in 1998. In the 1994-1999 period, the average and median of the planned budgets were around 6 thousand dollars and 4 thousand dollars, respectively.

---

[21] For simplicity, the authors assume that the interest rate is 0%.

[22] Here, the authors use flow concept, not stock concept.

[23] The purposes of planning are 1) installing new computer equipment and computer software, 2) saving labor, and 3) rationalization of office works, and so on.

Table 6.6. Statistics of the planned budgets concerning information systems in the 1993-1999 period

Year	Average*	Median*	SD	Minimum*	Maximum*	Number of banks
1993	3167.41	1116.00	5166.34	0	19000	32
	0.72	0.39	1.08	0.00	5.72	
1994	3057.44	1230.00	6675.47	0	36800	32
	0.57	0.44	0.53	0.00	2.10	
1995	5146.92	973.00	8050.77	0	37520	38
	0.68	0.42	0.93	0.00	5.36	
1996	4460.33	1070.00	7078.05	0	27520	39
	0.58	0.39	0.57	0.00	1.91	
1997	4359.55	995.50	6807.36	0	23136	38
	0.60	0.43	0.71	0.00	3.77	
1998	3788.26	859.00	5854.85	0	26626	43
	0.56	0.41	0.65	0.00	3.64	
1999	3841.93	777.00	10684.39	0	79300	61
	0.61	0.32	0.82	0.00	4.01	

Note that the lower in each row represents the system asset value per employee
* one million yen

6.6 Classification of Banks by Information System Assets

In the previous section, the authors reviewed information system assets, computer software assets, and computer equipment in the Japanese banking industry. However, we have the following questions: first of all, why not many banks write information system assets in the financial statements voluntary before 1999. Second, why not some banks write information system assets in the financial statements after implementing 1998 R&D standards?

The Authors claim a unique insight of these questions. The key is accounting rule so-called "1% rule" in Regulations of Financial Statements, Article 120. The 1% rule is based on the relative importance of assets and their accounting procedure. The principle of the significance in accounting is as follows: when firms provide information on the firm's financial status and the business performance, the firm must write the information in the financial statement. In addition, the items must also be charged correctly. On the contrary, the rule also permits to simplify the accounting procedures if it is small significant. Concretely, it is judged as small significance if the percentage of total assets is less than 1%. In our examination, many information system assets were lumped together and was written "other assets" in financial statements. The 1% rule causes that such information is not explicitly written in the financial statements. If a management executive bases his judgment on significance of information systems, especially computer software assets of the bank, or if the share of assets is less than 1% of the total asset, he can apply 1% rule to the computer software asset and the assets are not written in the financial statement.[24] Then, we cannot grasp exact information system assets of each bank if computer software assets and other assets are treated collectively. In fact, all banks can apply computer software asset to 1% rule in the 1993-1999 period, as wee see the share of computer software assets in total assets is about 0.19% in Subsect. 6.5.1.

The banks can decide whether or not they apply computer software assets to 1% rule. Here, we can image that a bank clearly writes computer software in the financial statement if the bank places high priority on computer software for resource of management. Otherwise, the bank would not write computer software in the financial statement. Therefore, we can also classify the Japanese banks into two groups[25]. One is a (bank) group places high priority on information system. The authors assume that this group satisfies (1) banks write the information concerning information system assets in the financial statements voluntary before implementing the 1998 R&D standards, or (2) banks write the information in the financial statements after implementing the 1998 R&D standards.[26] These banks in this group place high priority on information system for resource of management. On the other hand, another is and (bank) group places low priority on information system. This group consist of banks avoid disclosing the information, and they place low priority on information system for resource of management. In Chap. 9, the authors will analyze the effects of information system investment and computer software investment in the Japanese banking industry by using balanced and unbalanced panel data. Each panel data is one we described in this chapter.

It is a key for the second question we guess that some banks did not have enough preparation periods for applying accounting procedures on computer software in addition to applying 1% rule.

---

[24]Some banks mention clearly that they regard "other assets" as computer software in the financial statements. Note that in this case, the authors recognize "other assets" as computer software of the banks.

[25]Takemura (2002) includes some variance analyses between two groups. As a result, Takemura (2002) reports significant differences between two groups in some items.

[26]However, after implementing the 1998 R&D standards, it became too difficult to distinguish between voluntary disclosure and legal disclosure.

6.7 Summary

In Chap. 6, the authors have described the circumstances concerning information system assets in the Japanese banking industry. As a result, the authors find that the number of banks writing the information about information system in the financial statements increased by implementing the 1998 R&D standards. In addition, the authors viewed the accounting procedure on information systems was vastly improved. However, the proportion of banks is only 55% despite of implementing the 1998 R&D standards and the practical guidelines on computer software.

Finally, note that our description may be hardly adequate (insufficient) because there is not enough data. In the previous chapter, the authors define information system assets; Information System Asset I and II. The authors also emphasized on the need for non-financial information related-information system such as human recourses, type of organization, and consumer satisfaction. But such information is not written in the financial statements in spite of the significant management as we viewed in this chapter and Chap. 2. Especially, the authors described significance of information system as management strategy in Chap. 2.

Therefore, the authors should discuss the feasibility of "IT accounting" minutely.[27] We image that IT accounting rule requires both financial and non-financial information related-information system are written in the financial statements. We expect to establish IT accounting along with more improved of accounting standards on information system. If IT accounting is established, our research would be advanced more.

---

[27]We will mention the possibility of balanced score card (BSC) in Chap. 9.

Cross-Section Analysis of Information System Investment

Y. Ukai, and T. Takemura

7.1 Introduction

The purpose of this chapter is to analyze information system investment of the banking industry in Japan by using the micro data from the original questionnaire surveys and by the cross-section in each period.[1] Until the latter half of the 1990s, information system investment was comprehended only at an aggregate level from industrial sources in Japan.

As mentioned in Chap. 5, questionnaire surveys were implemented four times. The purpose of the survey was to clarify the relationship between the influences of IT investment on bank behavior and their achievements. The authors analyzed the relationships by connecting the questionnaires and the micro data from financial statements or management information that was disclosed as stipulated under Article 21 of the Banking Law in Japan.[2]

It should be noted that the authors cannot specify the name or location of each bank because of their agreement with the banks that provided the

---

[1]The Bank of Japan officially announces the aggregated figures about information system investment in about 200 financial companies (banks, securities, insurance) in Short-Term Economic Survey of Principal Enterprise in Japan, in May of each year. This source reported that the annual total amount for information system investment peaked in 1991 at 1500 billion yen, decreased to 1400 billion yen in 1992, and remained unchanged at 1100 billion yen until 1995. However, these micro data were disclosed only to researchers belonging to the Bank of Japan.

[2]Article 21 of Banking Law: Banks must prepare a report that states the situation of the business and the assets, and provide it publicly by submission at the main branch office in every year of operation. However, the items that may erode confidence, harm the confidentiality of depositors and other traders, place individuals at a disadvantage with respect to the business conducted in banks, or excessively impose a burden should be excluded.

useful information necessary for their research. In addition, the authors do not disclose the distribution which can be estimated from its information.

7.1.1 Definition of Information System Investment and Questionnaire Items

In this chapter, the authors identify information system investment of the banking industry as "Information System Investment I," which was already defined in Chap. 5. The authors define information system investment as: (1) mainframes, workstations, personnel computers, and cash dispenser (CD), and the cost of equipment and/or rental for terminal devices such including CDs and automatic teller machines (ATMs), (2) purchase fees and charges for computer software, and (3) personnel expenses related to (1) and (2).

On the last day of February 1995, the authors mailed the first questionnaire to the directors of the information system departments in 120 banks, which were listed on eight security exchanges in Japan. The authors collected completed questionnaires by the last day of April 1995 from 26 banks, the response rate being 21.6%.

The authors mailed the second questionnaire to the directors of the same 120 banks that received the first questionnaire at the end of February 1996. The authors collected 30 completed questionnaires by the end of April 1996, the response rate being 25%.

The authors mailed the third questionnaire to the directors of 119 banks in December 1997. The Hokkaido Takushoku Bank was excluded on this occasion due to its being declared bankrupt immediately before the questionnaire period. The authors collected 22 completed questionnaires by the end of April 1998, the response rate being 18.5%.

The authors mailed the fourth questionnaire to the directors of 117 banks in March 2000. Further 2 banks were excluded from the questionnaire after having undergone mergers. The authors collected six completed questionnaires by the end of May 2000, the response rate having decreased sharply to 5%. (See Chap. 5 for a discussion of this trend.)

In this chapter, the results from the first three questionnaires are statistically analyzed, with the results of the fourth questionnaire being excluded. The questions of the first questionnaire concerned the following 12 items:

1. Operative start time of the information system.
2. The development time of the information system.
3. Information system development personnel and their position breakdown.
4. Development costs of the information system.
5. Annual income of development personnel.
6. Outsourcing situation of development costs for the information system.
7. System operation personnel and their position breakdown.
8. Operation costs of the information system.
9. Annual income of information system operation personnel.

10. Outsourcing situation of operation cost for the information system.
11. Renewal time for the information system.
12. Investment standard in case of information system renewal.

The question topics in the second questionnaire were essentially the same as those in the first. One extra topic (discussed in detail in Chap. 5), was also added:

13. The confusion of the information system investment concept.

The third questionnaire became more complicated than the first two because the authors asked for more detail than in the earlier questionnaires, and sought information on how the system was dividing into the accounting system, and narrowly-defined information system and other main systems. Furthermore, to understand aspects of management structure, new questions were also posed concerning the formation of special teams to combat the effects of financial deregulation and the Y2K issue.

7.1.2 Financial Index and Total Employee Number in Answering Banks

The authors investigated the important values in the 26 banks of the first questionnaire based on the accounting in the 1994 fiscal year, by using the "Financial Statements Overview" of each bank which was published by the Printing Bureau in the Ministry of Finance at that time.

The average of the total assets was about 2900 billion yen. The maximum was about 17 trillion yen, the minimum was about 300 billion yen, and the median was about 1900 billion yen. This implies that there is a quite significant number of small-scale banks. The average of the loan and bills discounted was about 2 trillion yen, and the maximum and minimum were 10600 billion yen and 205 billion yen, respectively. The median was about 1250 billion yen. The average of the credit balance was about 2600 billion yen, and the maximum and minimum were about 17700 billion yen and 125 billion yen, respectively. The median was about 1250 billion yen. For employee numbers in 1994, the average number was about 2400, and the maximum and minimum were about 9600 and 500, respectively. The median was about 1900.

The authors found that most of the respondents were regional banks and the second regional banks (i.e., smaller banks than the regional banks). Therefore, it was possible to find some statistically significant conclusions by using the first questionnaire.

The authors investigated the important data of the 30 banks that responded to the second questionnaire based on the accounting in the 1995 fiscal year, by using each Financial Statements Overview. The average total assets was about 7700 billion yen. The maximum was about 53 trillion yen, and the minimum was about 310 billion yen. In addition, the median was about 1980 billion yen. This implies that nationwide banks were included in the respondents. The average of the loan and bills discounted was about 5 trillion yen, and the maximum and minimum were 35500 billion yen and 220 billion yen, respectively. The median was 1360 billion yen. The average credit balance was about 5500 billion yen, and the maximum and minimum balances were about 35500 billion yen and 285 billion yen, respectively. The median was 1750 billion yen. The average total employee number in 1995 was about 3700, the maximum and minimum numbers were about 18000 and 500, respectively, and the median was about 2100.

The authors found that the respondents included not only the regional banks and the second regional banks, but also nationwide banks. Therefore, it was possible for some statistically significant conclusions to be drawn from the results of the second questionnaire.

The authors investigated the important data of the 22 banks that responded to the third questionnaire based on the accounting in the 1997 fiscal year, by using each Financial Statements Overview. The average total assets was about 6600 billion yen. The maximum was about 47 trillion yen, and the minimum was about 340 billion yen. The median was about 1700 billion yen. These data imply that nationwide banks were also included in the banks that answered this questionnaire. The average of the loan and bills discounted was about 4700 billion yen, and the maximum and minimum were 35 trillion yen and 240 billion yen, respectively. The median was 1200 billion yen. The average credit balance was about 4400 billion yen, and the maximum and minimum were about 33 trillion yen and 290 billion yen, respectively. The median was about 1500 billion yen. The average total number of employees in 1997 was about 3350, the maximum and minimum were about 17 400 and 500, respectively, and the median was about 1700.

The respondents to the third questionnaire included regional banks, second regional banks, and also nationwide banks. Therefore, it was possible for some statistically significant conclusions to be drawn from the results of the third questionnaire. However, in comparing the second questionnaire with the third one, various statistics vividly showed slackening growth in the whole banking industry in Japan.

7.2 Various Statistics of Questionnaires

Operation Starting Year of Current Information System

In the first questionnaire, 25 valid answers were received for the operation starting year of the current information system. The average operation starting year was 1988, the median was 1990, and most of the banks had started by 1993. As mentioned in detail at Chap. 1, this year corresponds to the period when the information systems of some of the major nationwide banks were divided into independent subsystems. For example, a basic accounting

Table 7.1. Statistics from answers to the first questionnaire

	Average	SD	Median	Kurtosis	Skewness
Operation starting year of current information system	1988	6.58	1990	3.66	−1.91
Development starting year of current information system	3.2	1.19	3	0.10	0.53
Total labor input for current information system development[a]	8373.7	9.6×10^3	7200	6.52	2.36
Regular labor input for current information system development[a]	2281.9	2.0×10^3	1800	0.91	1.21
Keiretsu subsidiary labor input for current information system development[a]	1489.2	2.0×10^3	576	0.89	1.46
Internal system development cost[b]	103.2	1.7×10^2	40	10.51	3.02
Average annual income of system development personnel[c]	605.7	1.3×10^2	600	0.55	0.17
Average age of system development personnel[d]	—	—	—	—	—
System development expense for keiretsu subsidiaries[b]	4.4	9.5×10^4	0.4435	7.20	2.71
System development expense for non keiretsu outsourcing[b]	13.8	1.9×10^5	4.9	0.46	1.31
Total labor input for current information system operation[a]	539.7	7.4×10^2	270	4.17	2.29
Regular labor input for current information system operation[a]	243.4	3.2×10^2	120	4.58	2.31
Keiretsu subsidiary labor input for current information system operation[a]	87.7	1.3×10^2	42	8.54	2.60
System operation cost without personnel cost[b]	37.0	4.5×10^5	1.4	2.78	1.82
Average annual income of system operation regular personnel[c]	525.7	1.1×10^2	500	−0.74	0.01
System operation expense for keiretsu subsidiaries[b]	1.0	2.6×10^4	1100	15.21	3.81
System operation expense for non keiretsu outsourcing[b]	1.5	2.7×10^4	2400	4.42	2.26
Time between information system renewals[d]	5.6	3.10	5	−1.29	0.21

a: man months, b: 100 million yen, c: 10 thousand yen, d: years

Table 7.2. Statistics from answers to the second questionnaire

	Average	SD	Median	Kurtosis	Skewness
Operation starting year of current information system	1990	4.15	1991.5	−0.18	−0.86
Development starting year of current information system	3.5	1.28	3	0.82	0.92
Total labor input for current information system development[a]	9051.4	1.48×10^4	3910	7.61	2.77
Regular labor input for current information system development[a]	2793.0	$4.28 times 10^3$	1488	14.41	3.53
Keiretsu subsidiary labor input for current information system development[a]	2462.8	8.19×10^3	305	24.25	4.86
Internal system development cost[b]	172.1	2.57×10^2	60	2.96	1.94
Average annual income of system development personnel[c]	641.8	1.58×10^2	619	0.45	−0.24
Average age of system development personnel[d]	33.3	3.79	35	−0.45	−0.52
System development expense for keiretsu subsidiaries[b]	14.5	3.90×10^5	30	7.39	2.90
System development expense for non keiretsu outsourcing[b]	26.7	4.82×10^5	6.1	11.59	3.22
Total labor input for current information system operation[a]	932.4	1.42×10^3	282	5.17	2.30
Regular labor input for current information system operation[a]	286.3	3.40×10^2	180	1.90	1.67
Keiretsu subsidiary labor input for current information system operation[a]	554.8	2.21×10^3	24	27.41	5.17
System operation cost without personnel cost[b]	35.6	4.66×10^5	19.8	2.76	1.89
Average annual income of system operation regular personnel[c]	575.8	1.79×10^2	606	4.79	−1.87
System operation expense for keiretsu subsidiaries[b]	7.4	3.32×10^5	0.135	25.75	5.06
System operation expense for non keiretsu outsourcing[b]	1.5	2.90×10^4	0.36	4.39	2.37
Time between information system renewals[d]	5.6	4.58	5	−0.65	0.50

a: man months, b: 100 million yen, c: 10 thousand yen, d: years

Table 7.3. Statistics from answers to the third questionnaire

	Average	SD	Median	Kurtosis	Skewness
Operation starting year of current information system	1991	3.73	1991.5	0.87	-0.88
Development starting year of current information system	3.2	1.31	3	2.11	1.26
Total labor input for current information system development[a]	4002.1	4.77×10^3	2560	6.02	2.21
Regular labor input for current information system development[a]	1209.6	1.47×10^3	940	9.80	2.80
Keiretsu subsidiary labor input for current information system development[a]	1161.3	2.26×10^3	37	5.31	2.41
Internal system development cost[b]	382.1	8.80×10^2	90	11.48	3.32
Average annual income of system development personnel[c]	667.9	1.32×10^2	600	0.50	1.19
Average age of system development personnel[d]	34.3	3.00	35	2.46	0.46
System development expense for keiretsu subsidiaries[b]	17.1	3.28×10^5	0	6.51	2.50
System development expense for non keiretsu outsourcing[b]	78.4	1.64×10^6	7.9	5.40	2.37
Total labor input for current information system operation[a]	1050.3	1.78×10^3	336	8.30	2.85
Regular labor input for current information system operation[a]	493.3	9.86×10^2	186	15.44	3.83
Keiretsu subsidiary labor input for current information system operation[a]	457.3	8.55×10^2	84	4.17	2.26
System operation cost without personnel cost[b]	39.4	5.76×10^5	20	3.11	2.05
Average annual income of system operation regular personnel[c]	648.5	1.16×10^2	600	0.62	0.76
System operation expense for keiretsu subsidiaries[b]	1.1	3.15×10^4	0.675	13.33	3.62
System operation expense for non keiretsu outsourcing[b]	0.5	5.87×10^3	0.13	0.81	1.26
Time between information system renewals[d]	4.2	3.87	3	-0.82	0.90

The accounting system was mentioned only in the third questionnaire

a: man months, b: 100 million yen, c: 10 thousand yen, d: years

system may have been divided into a financial resources credit bill system, an international system, and so on.

The first information system started operation in 1969, and the most recent system commenced operation in 1995. Generally, the small-scale regional banks tended to use an identical information system over long periods.

In the second questionnaire, 28 valid answers were received for the operation starting year of the current information system. The average operation starting year was 1990, the median was 1991, and most of the banks had started by 1993, as in the first questionnaire. The oldest information system had started operation in 1981, and the most recent system commenced operation in 1996. Information systems in most banks tended to be renewed entirely rather than in parts.

In the third questionnaire, the authors subdivided the information system for their analysis. The number of banks that answered the operation starting year of the accounting system was 22. The average operation starting year was 1991, the median was 1991, and most banks had started by 1995. The oldest information system was started in 1981, and the newest system was started in 1996. As a whole, narrowly-defined information systems in most banks tended to become newer. There are few differences in the statistics between the accounting systems and the narrowly-defined information systems. However, it was in 1985 that the oldest narrowly-defined information system started operation. It was found that the development of narrowly-defined information systems was generally achieved after establishing the accounting system. However, for other systems, significant trends could not be identified because the answers were spread. The development of most systems in banks had finished by the mid-1990s, and development of the newest system finished immediately prior to the third questionnaire.

Required Development Time for Current Information System

In the first questionnaire, 24 valid answers were received for the required development time for the current information system. The average development time was 39 months, the median was 3 years, and most of the banks answered 3 years. The longest required development time was 6 years, and the shortest was 1 year.

In the second questionnaire, 28 valid answers were received for the required development time for the current information system. The average development time was 42 months, the median was 3 years, and most of the banks answered 3 years. The longest development time was 7 years, and the shortest was 18 months. The development times were found to be longer in the second questionnaire when compared with the first.

In the third questionnaire, 22 valid answers were received for the required development time for the accounting system. The average development time was 38 months, the median was 3 years, and most of the banks answered

3 years. The longest required development time was 7 years, and the shortest was 18 months. Twenty-one valid answers were received for the required development time of narrowly-defined information systems. The average development time was 2 years, the median was 2 years, and most of the banks answered 1 year. The longest required development time was 7 years, and the shortest was 1 year. Overall, the development times for the narrowly-defined information systems were shorter than those for the accounting systems. For other systems, no significant trends could be identified due to data spread.

Total Labor Input for Current Information System Operation

The information system development personnel for each bank include the head office regular employees, the keiretsu subsidiary employees, and the employees of the non-keiretsu outside bank. Through all questionnaires, "man months" was used as the unit to measure the labor input. In addition, the authors obtained clarification from the information system development persons in charge by telephone interview when the answers were ambiguous.

In the first questionnaire, 23 valid answers were received for the total labor input for development of the current information system. The average labor input was 8400 man months, and the median was 7200 man months. The maximum and minimum inputs were 42000 man months and 200 man months, respectively. Considerable differences were found in the responses of the banks that answered.

In the first questionnaire, 22 valid answers were received for the regular labor input for development of the current information system. The average labor input was 2280 man months, and the median was 1650 man months. The maximum and minimum inputs were 7300 man months and 100 man months, respectively. Considerable differences were also found in the regular labor input for the information system between banks.

In the first questionnaire, 21 valid answers were received for the keiretsu subsidiary labor for development of the current information system. The average labor input was 1500 man months, and the median was 576 man months. The maximum and minimum inputs were 6700 man months and 0 man months, respectively. Considerable differences were also found between banks for the keiretsu subsidiary labor input for information systems.

It should be noted that several regional banks commented that it was not possible to comprehend the correct number of the system personnel because of the joint development of information systems. Curiously, there were many cases in which joint development was conducted between the distant regional banks.[3] In this case, the authors divided the total labor input by the number of banks that participated in the joint development.

---

[3] Joint Stock Corporation Kanto Data Center was invested in by three regional banks in December 1977, and is an example of the joint development and joint operation of an information system. This company was delegated business by five regional banks in May 1991.

In the second questionnaire, 28 valid answers were received for the total labor input for the development of the current information system. The average labor input was 9050 man months, and the median was 3900 man months. The maximum and minimum inputs were 63000 man months and 50 man months, respectively. The very large maximum input implies that the human investment required for system development in nationwide mega banks was very large, as mentioned in Chap. 1.

In the second questionnaire, 25 valid answers were received for the regular labor input for development of the current information system. The average labor input was 2800 man months, and the median was 1500 man months. The maximum and minimum inputs were 21000 man months and 45 man months, respectively. Considerable differences were found between banks for the regular labor input for information systems, as was observed in the first questionnaire.

In the second questionnaire, 26 valid answers were received for the keiretsu subsidiary labor input for development of the current information system. The average labor input was 2460 man months, and the median was 305 man months. The maximum and minimum inputs were 42000 man months and 0 man months, respectively. Considerable differences were also observed between banks for the keiretsu subsidiary labor input for information systems. As observed above, the labor input for information system development became much larger in the second questionnaire, which included most nationwide banks.

In the third questionnaire, the authors subdivided information systems for their analysis. Twenty valid answers were received for the total labor input for development of the accounting system. The average labor input was 4000 man months, and the median was 2600 man months. The maximum and minimum inputs were 20000 man months and 50 man months, respectively. Similar trends to those observed in the second questionnaire were also observed.

In the third questionnaire, 19 valid answers were received for the regular labor input for the development of the accounting system. The average labor input was 1200 man months, and the median was 940 man months. The maximum and minimum inputs were 6500 man months and 30 man months, respectively. Considerable differences were observed between banks for the regular labor input for accounting system development.

In the third questionnaire, 18 valid answers were received for the keiretsu subsidiary labor for development of the accounting system. The average labor input was 1160 man months, and the median was 40 man months. The maximum and minimum inputs were 8000 man months and 0 man months, respectively. The authors found considerable differences between banks for the keiretsu subsidiary labor for the accounting system.

For the development of the narrowly-defined information system, 18 valid answers were received for the total labor input. The average labor input was 810 man months, and the median was 250 man months. The maximum and minimum inputs were 6200 man months and 5 man months, respectively.

Seventeen valid answers were received for the regular labor input of the narrowly-defined information system development in the third questionnaire. The average labor input was 360 man months, and the median was 120 man months. The maximum and minimum inputs were 2340 man months and 3 man months, respectively. Considerable differences were observed between banks for the regular labor input for development of the narrowly-defined information systems.

Sixteen valid answers were received in the third questionnaire for the keiretsu subsidiary labor input for development of the narrowly-defined information system. The average labor input was 90 man months, and the median was 2 man months. The maximum and minimum inputs were 650 man months and 0 man months, respectively. Considerable differences were also observed between banks for the keiretsu subsidiary labor input for development of the narrowly-defined information systems. For the other systems, no significant trends could be observed because of data spread.

The labor input for each system development was of small scale and required 600-800 man months on average. However, for a nationwide bank that developed strategically, the bank's labor input was about 3200 man months. This situation is an example of the system development strategies in nationwide mega banks in Japan as mentioned in Chaps. 1 and 2.

Internal System Development Cost

In the first questionnaire, 20 valid answers were received for the internal system development costs, excluding personnel expenses and outsourcing expenses. It should be noted that in cases of joint development, unless the exact share of expenses in each bank was made clear in the questionnaire or from telephone interview, the total development cost was equally divided by the number of cooperating banks. The average system development cost was 10 billion yen, and the median was 4 billion yen. Four banks reported that the system development cost was either zero or less than 100 million yen. The maximum and minimum costs were 75 billion yen and 0 yen, respectively. In addition, six banks reported that the internal system development cost was more than 10 billion yen.

In the second questionnaire, 23 valid answers were received for the internal system development cost, excluding personnel expenses and outsourcing expenses. The average system development cost was 17 billion yen, and the median was 6 billion yen. The mode was 50 billion yen. The maximum and minimum costs were 90 billion yen and 0 yen, respectively. This implied that the nationwide banks were represented in the results.

For the third questionnaire, the system development cost was restricted to include the development of the third-generation on-line system and after that. The number of valid banks was 15. The average and median costs were about 38 billion yen and 9 billion yen, respectively. The mode was 25 billion yen, and the maximum and minimum costs were 340 billion yen and 500 million yen, respectively.

Average Annual Income and Age of System Operation Personnel

In the first questionnaire, 18 valid answers were received for the average annual income of system operation personnel. The average and median incomes were 6.05 million yen and 6 million yen, respectively. The mode was also 6 million yen. The maximum and minimum incomes were 9 million yen and 3.5 million yen, respectively.

Comparing the results with the annual incomes of employees in nation-wide banks reported in newspapers and economic magazines at that time, the reported incomes were much lower. This implied that almost all the respondents were regional banks. One of the reasons for this may have been because the job evaluation-based wage system had not been implemented in Japan at that time.

In the second questionnaire, 21 valid answers were received for the average annual income of system operation personnel. The average and median incomes were 6.4 million yen and 6.2 million yen, respectively. The mode was 7 million yen. The maximum and minimum incomes were 9.5 million yen and 3 million yen, respectively. The range in annual income was larger than that in the first questionnaire. Respondents were also asked to report the average age of system operation personnel. The average age was 33 years, the median was 35 years, and the highest and lowest average ages were 40 years and 26 years, respectively.

In the third questionnaire, 14 valid answers were received for the average annual income of system operation personnel. The average and median incomes were 6.7 million yen and 6 million yen, respectively. The mode was 6 million yen. The maximum and minimum incomes were 9.5 million yen and 5 million yen, respectively. The average age was 34 years, the median was 35 years, and the highest and lowest average ages were 42 years and 28 years, respectively.

System Development Outsourcing Costs

In the first questionnaire, 18 valid answers were received for the system development outsourcing costs of keiretsu subsidiaries. The average and median costs were 400 million yen and 44 million yen, respectively. The mode and maximum costs were nil and 3500 million yen, respectively.

In the first questionnaire, 18 valid answers were received for the system development outsourcing costs of companies without close capital ties (non-keiretsu). The average and median costs were 1400 million yen and 500 million yen, respectively. The mode and maximum costs were nil and 5800 million yen, respectively.

Total outsourcing costs were calculated by summing the keiretsu and non-keiretsu costs for each bank. The average and median costs were 2100 million yen and 1 billion yen, respectively. The maximum cost was 7100 million yen.

Considering the above three types of distribution with the annual income distribution of the development personnel, it appears that there was demand in the order of hundreds of billions of yen for information system development in regional banks. However, the engineer labor market in the banking industry did not emerge because of significant differences in the annual income of the development personnel among banks and this gap did not shrink.

On the contrary, nationwide mega banks might become suppliers if they supply their knowledge and modes of system development that they have accumulated.[4]

In the second questionnaire, 21 valid answers were received for the system development outsourcing costs of the keiretsu subsidiary. The average and median costs were 1400 million yen and 0.3 million yen, respectively. The mode and the maximum costs were nil and 14 billion yen, respectively. Seventeen valid answers for the system development outsourcing costs of non keiretsu companies. The average and median costs were 2700 million yen and 600 million yen, respectively. The mode and the maximum costs were 40 million yen and 20 billion yen, respectively.

In the third questionnaire, nine valid answers were received for the system development outsourcing costs of keiretsu subsidiaries. The average costs were 1700 million yen. The mode and the maximum costs were nil and 10 billion yen, respectively. Ten valid answers were received for the system development outsourcing costs of non keiretsu companies. The average and median costs were 7800 million yen and 800 million yen, respectively. The maximum cost was 50 billion yen. The authors found that the outsourcing of system development became more extensive in some banks.

Number and Breakdown of Information System Operation Personnel

In the first questionnaire, 26 valid answers were received for the total labor input for system operation. The average labor input was 540 man months, and the median was 270 man months. The mode was 240 man months. The maximum and minimum inputs were 2760 man months and 36 man months, respectively. Considerable differences were observed between the banks for the labor input to information system operation.

---

[4]The mega banks established information system development organizations as related companies. These capital relationships were complicated, and because their form of organization was often as a joint company between the subsidiary and the sub-subsidiary, they were often undisclosed. However, human relationships were closely maintained by the personnel transfer system. In the past, these nationwide bank-related companies had non-bank customers. However, there were several examples that had regional banks as customers.

In the first questionnaire, 25 valid answers were received for the regular labor input to information system operation. The average labor input was 240 man months, and the median was 120 man months. The mode was 60 man months. The maximum and minimum labor inputs were 1200 man months and 2 man months, respectively. Considerable differences were observed between banks for regular labor input to information system operation.

In the first questionnaire, 24 valid answers were received for the keiretsu subsidiary labor input to information system operation. The average labor input was 90 man months, and the median was 40 man months. The maximum and minimum labor inputs were 560 man months and nil, respectively.

The authors extracted the number of the regular employees from the Financial Statements Overview of each bank, and calculated the system operation regular employee ratio of 25 banks. The ratios of 21 banks were uniformly distributed and less than 1%. The ratios in 4 banks exceeded more than 2%. The average, median, maximum, and minimum ratios were 0.9%, 0.6%, 4.2%, and 0.01%.

In the second questionnaire, 30 valid answers were received for the total labor input to information system operation. The average labor input was about 930 man months, and the median was 280 man months. The mode was 240 man months. The maximum and minimum labor inputs were 6000 man months and nil, respectively. The authors found considerable differences between banks for the regular labor input to information system operation.

In the second questionnaire, 29 valid answers were received for the regular labor input to the information system operation. The average labor input was 290 man months, and the median was 180 man months. The mode was 0 man months, and the maximum and minimum inputs were 1200 man months and nil, respectively.

In the second questionnaire, 30 valid answers were received for the keiretsu subsidiary labor input to information system operation. The average labor input was 550 man months, and the median was 24 man months. The mode was 0 man months, and the maximum and minimum inputs were 12000 man months and nil, respectively.

In the third questionnaire, 19 valid answers were received for the total labor input to accounting system operation. The average labor input was about 1050 man months, and the median was 340 man months. The mode was 170 man months. The maximum and minimum inputs were 7200 man months and 130 man months, respectively. Eighteen valid answers were received for the regular labor input to accounting system operation. The average labor input was 490 man months, and the median was 190 man months. The mode was 0 man months. The maximum and minimum inputs were 4300 man months and nil, respectively. Eighteen valid answers were also received for the keiretsu subsidiary labor input to accounting system operation. The average labor input was 460 man months, and the median was 84 man months. The mode was 0 man months. The maximum and minimum inputs were 2900 man months and nil, respectively.

System Operation Costs Excluding Personnel Costs

In the first questionnaire, the question concerning the total annual cost for current system operation including rental, lease, and running costs, but excluding regular personnel costs, received 20 valid answers. The average and median costs were 3700 million yen, and 1400 million yen, respectively. The mode was 1400 million yen. The maximum and minimum costs were 17 billion yen and 0.2 million yen, respectively.

In the second questionnaire, the question concerning the total annual cost for current system operation including rental, lease, and running costs, but excluding regular personnel costs, received 23 valid answers. The average and median costs were 3600 million yen, and 2 billion yen, respectively. The mode was 500 million yen. The maximum and minimum costs were 17 billion yen and 30 million yen, respectively.

In the third questionnaire, 15 valid answers were received for the annual total cost, excluding costs of regular personnel, for operation of the current system. The average and median costs were 3900 million yen, and 2 billion yen, respectively. The mode was 1 billion yen. The maximum and minimum costs were 18 billion yen and 120 million yen, respectively.

From the data, it appears that there was a demand in the order of billions of dollar for information system operation in regional banks, much the same as described above.

Annual Income of Regular System Operation Personnel

In the first questionnaire, the question concerning the annual income of regular system operation personnel received 17 valid answers. The average and median incomes were 5.25 million yen, and 5 million yen, respectively. The mode was 6 million yen. The maximum and minimum incomes were 7.3 million yen and 3.5 million yen, respectively. The average annual income of the system operation personnel was about 1 million yen lower than that of the system development personnel. This was probably because the length of the service year for operation personnel was shorter, and development personnel received significant shift allowance because of the three-shift work roster as described in Chap. 1.

In the second questionnaire, 20 valid answers were received for the annual income of regular personnel conducting system operation. The average and median incomes were 5.75 million yen, and 6.1 million yen, respectively. The mode was 6 million yen. The maximum and minimum incomes were 8 million yen and 3.5 million yen, respectively.

In the third questionnaire, 13 valid answers were received for the annual income of regular operation personnel. The average and median incomes were 6.5 million yen, and 6 million yen, respectively. The mode was 6 million yen. The maximum and minimum incomes were 9 million yen and 4.8 million yen, respectively.

System Operation Outsourcing Costs and Cost Breakdown

In the first questionnaire, the question concerning the system operation outsourcing costs for keiretsu subsidiaries received 18 valid answers. The average and median costs were 100 million yen and 11 million yen, respectively. The mode was nil. The maximum and minimum costs were 1100 million yen and nil, respectively. Nineteen valid answers were received for the system operation outsourcing costs for non keiretsu companies. The average and median costs were 150 million yen and 24 million yen, respectively. The mode was nil. The maximum and minimum costs were 950 million yen and nil, respectively.

In the second questionnaire, 26 valid answers were received for the system operation outsourcing costs for keiretsu subsidiaries. The average and median costs were 700 million yen, and 1350 million yen, respectively. The mode was nil. The maximum and minimum costs were 17 billion yen and nil, respectively. Twenty-five valid answers were received for the system operation outsourcing costs for non keiretsu companies. The average and median costs were 150 million yen and 36 million yen, respectively. The mode was nil. The maximum and minimum costs were 1 billion yen and nil, respectively.

In the third questionnaire, 14 valid answers were received for the system operation outsourcing costs for keiretsu subsidiaries. The average and median costs were 100 million yen and 6.75 million yen, respectively. The mode was nil. The maximum and minimum costs were 1200 million yen and nil, respectively. Thirteen valid answers were received for the system operation outsourcing costs for non keiretsu companies. The average and median costs were 45 million yen and 13 million yen, respectively. The mode was nil. The maximum and minimum costs were 180 million yen and nil, respectively.

Renewal Plans for Information Systems

In the first questionnaire, 14 valid answers were received for the plans concerning information system renewal. The average and median planned times between renewals were 5.5 years and 5 years, respectively. The mode was 3 years.

In the second questionnaire, 15 valid answers were received for the plans concerning information system renewal. The average and median planned times between renewals were 5.5 years and 5 years, respectively. The mode was 10 years.

In the third questionnaire, 15 valid answers were also received for the plans concerning information system renewal. The average and median planned times between renewals were 4 years and 3 years, respectively. The mode was 4 years.

Critical Factors in Case of Information System Renewal

In the first questionnaire, the authors asked the chief information officer (or system development officer) to rate the importance of factors that influenced the plans for renewal of the current information system. Four alternatives were provided:

1. Technical obsolescence
2. Investment actions of the other banks
3. New product developments
4. Administrative guidance of the Bank of Japan and Ministry of Finance

Respondents were instructed to rank the level of importance of each factor by using 4 for the most important, 3 for the second-most important, and so on, with 1 for the least important. About 30% of banks selected technical obsolescence as the most important and another 30% selected new product developments. As a result, the weights of both these factors were about 36%.[5]

Estimation of Information System Sector Personnel Ratio

The numbers of information system development personnel and information system operation personnel from the first questionnaire were used to estimate the information system sector personnel ratio by dividing them into the regular employees of each bank on the Financial Statement Overview. The ratio was obtained for about 21 banks.

As the result, the average, median, maximum, and minimum values were 3%, 2.6%, 6.6%, and 0.04%. According to the Center for Financial Information Systems (1996), the computer development sector personnel ratio was 2.9% in nationwide banks, 2.8% in regional banks, and 2.1% in the second regional banks in 1995. Their questionnaire covered these figures from the aspect of micro data.[6]

7.3 Results of Regression Analysis

System development cost will be treated as some kind of investment on the knowledge asset belonging to particular firm. This concept was described in Sect. 5.4. The marginal system development cost with respect to output should be estimated. The relationship between information system development and number of personnel should be also estimated. The authors estimated regression curves by the least-squares method.

---

[5]In interviews from 1994 to 1998 with information system development persons in charge in Japanese mega banks, it was a common opinion that "there was no necessity to be instructed by the Bank of Japan or Ministry of Finance about information system technology."

[6]See Financial Information System Center (1996) pp432-433. The computer-related sector in this questionnaire is the system concerning planning, development, and operation, and it was corresponded to "information system sector" in their questionnaire.

7.3.1 Information System Development Cost and Loan and Bills Discounted

The dependent variable is the information system development cost DC_i (100 million yen), which was aggregated with respect to each bank by the questionnaire. The independent variable is loan and bills discounted K_i (1 million yen). Note that the subscript i represents the questionnaire order. The number of valid observations in the first questionnaire was 20. The regression is as follows:

$$DC_1 = 0.006309K_1 - 35.76 \tag{7.1}$$

The t-value of the independent variable K_1 is 11.80. The coefficient of determination could be adjusted for the degrees of freedom; Adj. R^2 was 0.87. As the result of the t-test, the coefficient is statistically significant at the 1% level. That is, an increase of 1 million yen of loan and bills discounted led to an increase of 0.63 million yen in the information system development cost. This has interpretability of 87%.

The number of valid observations in the second questionnaire was 23. The regression is as follows:

$$DC_2 = 0.001832K_2 + 55.45 \tag{7.2}$$

The t-value of the independent variable K_2 is 6.27, and Adj. R^2 is 0.63. As the result of the t-test, the coefficient is statistically significant at the 1% level. That is, an increase of 1 million yen of loan and bills discounted led to an increase of 0.18 million yen in the information system development cost. This has interpretability of 63%.

The number of valid observations in the third questionnaire was 15. The regression is as follows:

$$DC_3 = 0.010706K_3 - 41.35 \tag{7.3}$$

The t-value of the independent variable K_3 is 31.80, and Adj. R^2 is 0.98. As the result of the t-test, the coefficient is statistically significant at the 1% level. That is, an increase of 1 million yen of loan and bills discounted led to an increase of 1.07 million yen in the information system development cost. This has interpretability of 98%.

7.3.2 Information System Development Cost and Total Assets

The dependent variable is the information system development cost DC_i (100 million yen), which was aggregated with respect to each bank by the questionnaire. The independent variable is total asset A_i (1 million yen). Note that the subscript i represents the questionnaire order.

The number of valid observations in the first questionnaire was 20. The regression is as follows:

$$DC_1 = 0.0041A_1 - 30.78 \tag{7.4}$$

The t-value of the independent variable A_1 is 13.53, and Adj. R^2 is 0.90. As the result of the t-test, the coefficient is statistically significant at the 1% level. That is, an increase of 1 million yen of total assets led to an increase of 0.41 million yen in the information system development cost. This has interpretability of 90%.

The number of valid observations in the second questionnaire was 23. The regression is as follows:

$$DC_2 = 0.0012A_2 + 56.52 \tag{7.5}$$

The t-value of the independent variable A_2 is 6.48, and Adj. R^2 is 0.65. As the result of the t-test, the coefficient is statistically significant at the 1% level. That is, an increase of 1 million yen in total assets led to an increase of 0.12 million yen in the information system development cost. This has interpretability of 65%.

The number of valid observations in the third questionnaire was 14. The regression is as follows:

$$DC_3 = 0.0073A_3 + 42.99 \tag{7.6}$$

The t-value of the independent variable is 33.81, and Adj R^2 is 0.98. As the result of the t-test, the coefficient is statistically significant at the 1% level. That is, an increase of 1 million yen in the total assets led to an increase of 0.73 million yen in information system development costs. This has interpretability of 98%

7.3.3 Information System Development Costs and Number of Personnel

The dependent variable is the information system development cost DC_i (100 million yen), which was aggregated for each bank from the questionnaire. The independent variable is the number of personnel L_i (1 person). Note that the subscript i represents the questionnaire order.

The number of valid observations in the first questionnaire was 20. The regression is as follows:

$$DC_1 = 0.076785L_1 - 96.17 \tag{7.7}$$

The t-value of the independent variable L_1 is 12.87, and Adj. R^2 is 0.89. As the result of the t-test, the coefficient is statistically significant at the 1% level. That is, an increase of number of personnel led to an increase of 7.68 million yen in the information system development cost. This has interpretability of 89%.

The number of valid observations in the second questionnaire was 23. The regression is as follows:

$$DC_2 = 0.044588L_2 - 19.08 \qquad (7.8)$$

The t-value of the independent variable L_2 is 8.28, and Adj. R^2 is 0.75. As the result of the t-test, the coefficient is statistically significant at the 1% level. That is, an increase in the number of personnel led to an increase of 4.46 million yen in the information system development cost. This has interpretability of 75%.

The number of valid observations in the third questionnaire was 15. The regression is as follows:

$$DC_3 = 0.227698L_3 - 309.44 \qquad (7.9)$$

The t-value of the independent variable L_3 is 13.57, and Adj. R^2 is 0.92. As the result of the t-test, the coefficient is statistically significant at the 1% level. That is, an increase in the number of personnel led to an increase of 22.76 million yen in the information system development cost. This has interpretability of 92%

7.4 Management Strategy and Investment Activity

Ukai (1997) conducted factor analysis with the information system investment ratio (information system development cost per information system employee) and the three factors (1) reaction to other banks, (2) new product developments, and (3) attitude to the Bank of Japan and Ministry of Finance. Banks were classified according to whether they emphasized the system factor renewal in their questionnaires, and 19 banks, whose system investment ratio was calculated, were analyzed. As a result, the reaction to the Bank of Japan and Ministry of Finance (dependent and independent types of strategy) was explained by the development investment ratio in banks, which was statistically significant at the 5% level. It was found that banks of the independent type developed information systems more positively than dependent banks.

The authors also checked that these results applied to the second and third questionnaires, by using dummy variables. The new dummy variable , which represents the management strategy, was added to Eqs. 7.1-7.6. The dummy variable is 0 if the banks accept the guidelines from the Bank of Japan and the government. Otherwise, the variable is 1.

The authors supposed that the dummy variable is a propensity enhancement variable concerning the information system investment. The multiple linear regression is as follows:

$$DC_i = aX_i + bD_iX_i + c \qquad (7.10)$$

where $X_i = K_i$, $X_i = A_i$, or $X_i = L_i$ are independent variables.

As for loan and bills discounted and total assets, the management strategy dummy variable D_3 was not statistically significant in the third questionnaire.

However, the management strategy dummy variable increased the investment propensity in the first and second questionnaires. Both were statistically significant at the 1% level, and the Adj. R^2 values were between 0.8 and 0.95.

$$DC_1 = \underset{(9.63)}{0.004473K_1} + \underset{(5.57)}{0.002822D_1K_1} - 26.95 \quad \text{Adj. } R^2 = 0.95 \quad (7.11)$$

$$DC_1 = \underset{(9.33)}{0.003045A_1} + \underset{(4.34)}{0.001517D_1A_1} - 22.31 \quad \text{Adj. } R^2 = 0.95 \quad (7.12)$$

$$DC_1 = \underset{(10.76)}{0.060664L_1} + \underset{(4.32)}{0.023663D_1L_1} - 86.43 \quad \text{Adj. } R^2 = 0.94 \quad (7.13)$$

$$DC_2 = \underset{(5.45)}{0.0013K_2} + \underset{(4.24)}{0.001726D_2K_2} + 40.32 \quad \text{Adj. } R^2 = 0.80 \quad (7.14)$$

$$DC_2 = \underset{(5.40)}{0.00088A_2} + \underset{(4.06)}{0.001063D_2A_2} + 44.61 \quad \text{Adj. } R^2 = 0.80 \quad (7.15)$$

$$DC_2 = \underset{(9.42)}{0.035129L_2} + \underset{(5.83)}{0.030898D_2L_2} - 47.16 \quad \text{Adj. } R^2 = 0.90 \quad (7.16)$$

Equations 7.1 and 7.11 were compared for the relation between the information system development cost and the loan and bills discounted. It was found that Adj. R^2 rises from 0.87 to 0.95. Moreover, propensity of investment increases from 0.006309 in Eq. 7.1 to 0.007295 if the dummy variable is 1 (the bank takes independent strategy).

Similarly, Eqs. 7.2 and 7.14 were compared and Adj. R^2 was found to rise from 0.63 to 0.80. The propensity of investment increased from 0.001832 in Eq. 7.2 to 0.003026 if the dummy variable is 1 (the bank takes independent strategy).

Equations 7.4 with 7.12 were compared for the relation between the information system development cost and the total assets. The Adj. R^2 value was found to rise from 0.90 to 0.95. The propensity of investment increased from 0.0041 in Eq. 7.4 to 0.004562 if the dummy variable is 1 (the bank takes independent strategy).

The Adj. R^2 values between Eqs.7.5 and 7.15 were also compared and the Adj. R^2 value was found to rise from 0.65 to 0.80. The propensity of investment increases from 0.0012 in Eq. 7.5 to 0.001943 if the dummy variable is 1. (the bank takes independent strategy)

The Adj. R^2 values between Eqs. 7.7 and 7.13 were compared for the relation between the information system development cost and the number of personnel. The Adj. R^2 value was found to rise from 0.89 to 0.94. However, propensity of investment decreased from 0.07678 in Eq. 7.7 to 0.003026 if the dummy variable is 1 (the bank takes independent strategy). That is, an increase in the number of personnel led to an increase in the system cost that was smaller in the independent banks than in the dependent banks.

On the other hand, Eqs. 7.8 and 7.16 were compared to find that Adj. R^2 rises from 0.89 to 0.94. The propensity of investment increased from 0.0445 in Eq. 7.8 to 0.066027 if the dummy variable is 1. (the bank takes independent strategy) That is, an increase in the number of personnel led to an increase of the system cost that was larger in the independent banks than in the dependent banks.

It is likely that the relation between the number of personnel and the propensity to invest for the information system was not affected by the management strategy variable. However, it is also likely that the relation between the system development cost and the loan and bills discounted, and the relation between the system development cost and the total assets had significant effects on the information system investment propensity.

In Chap. 8 the outsourcing strategy for information system investment is analyzed by using unbalanced panel data to determine whether the strategy impacts on the bank's market value.

7.5 Conclusions

Data observation and cross-section analysis reached the following statistical conclusions. First, the information system development cost had a positive correlation with the loan and bills discounted in each bank, and the total assets. Second, internal system development personnel and the system operation personnel were not always accorded the privileges of other sector personnel. Therefore, there were few possibilities for these personnel to move between banks through the labor market. Finally, banks with independent strategy, which do not take the guidelines from the Bank of Japan and governments into account, invest in information systems more positively accompanied with the increase of the loan and bills discounted and the total assets than banks of the dependent type.

8

Analysis of Information System Investment Using Questionnaire Data

S. Watanabe, Y. Ukai, and T. Takemura

8.1 Introduction

In the analysis of the production function in the United States banking by Prasad and Harker (1997), the effect of information technology capital on production was not statistically significant, as described in Chap. 4. The authors estimated the production function in Japanese banking by using the questionnaire data of information system asset. The earlier estimated result is not stable, and it is not possible to analyze the effect of information system assets using the result[1]. The specification problem of the banking product probably led to the unstable estimated result.[2]

The firm value approach reviewed in Chap. 4 is used in this chapter. This technique can grasp the effect of IT investment. The authors judged that the market value approach with both demand and supply side deserves careful attention, more so than the production function approach, which is limited to the supply side in the analysis of banking IT investment.

The term "information system assets" will be used instead of the term "information system capital" in this chapter to avoid confusion between capital

---

[1] Ukai and Kitano (2002), and Takemura (2003) analyzed the effect of information system assets in Japanese banking industry by using production function approach. They found the positive productivity of information system investment. See Sect. 4.2.

[2] Prasad and Harker (1997) used the amount of loan and deposit, profit, ROE (return on equity), and ROA (return on assets) as output of production functions of banking in the USA. It is possible to regard the deposit as output, because the deposit will produce implicit profit for the bank when the bank requires the deposit as a loan condition to the firm. In their study, a simple survey on the banking product was performed.

that is an accounting concept in financial statements and capital regarded as an economic concept as a production factor.

Earlier studies that have used the market value model include those of Brynjolfsson and Yang (1997) and Brynjolfsson, Hitt and Yang (2000, 2002). They analyzed the effects of computers and peripheral equipment on market value using the database of a private research agency that covered 820 nonfinancial companies in the USA over 8 years.

For the present study, the authors used the data on information systems in the Japanese banking industry as retrieved from questionnaires and interviews. Numeric data on information system investment in the banking industry were continuously collected in questionnaires from 1995. It is possible to use the results directly for econometric analysis, because the questions on IT investment in the questionnaire actually asked for numerical values. Both computer software investment and computer hardware investment were included in the data.

It is possible to obtain accurate amounts of information system investment, including the computer software associated with the computer hardware. This can avoid apparent overestimates for the effect of information system investment. In this chapter, the authors used the questionnaire data from 1995, 1996, and 1998. Over the period of the questionnaires in the 1995-1998 period on information system investment, every financial statement did not have the data of computer assets and computer software. The questionnaire survey was explained in detail in Chap. 5.[3]

The estimated result of the market value model explained in this chapter shows that information system assets of $1 in Japanese banking is relates to a market value of over $10. This is similar to result of analysis by Brynjolfsson and Yang (1997). They concluded that an increase of computer assets of $1 lead to an increase of $10 in the market value. They also indicated the existence of intangible assets, which they defined as computer software assets, accumulated education and training, and an organization reform effect, which is in sharp contrast to accounting terminology.[4] The amount of computer software assets, which were defined as intangible assets, were able to be determined from the questionnaires.

The authors made the outsourcing variable of information system development to analyze the organization reform effect, which Brynjolfsson, Hitt and Yang (2000, 2002) regarded as an intangible asset. The analysis also added the human resources of information system development to the information system assets.

---

[3]The Financial Institutions Center of Pennsylvania University Wharton School conducted a questionnaire on information system investment of the banking industry in the USA at about the same time. Prasad and Harker (1997) detailed their research results. See Sect. 4.2.

[4]Intangible fixed assets of banking are usually leasehold concession money and telephone subscriptions.

The composition of the remainder of this chapter is as follows. The model of multiple capitals with the adjustment cost by dynamic optimization model is shown and the data for the estimation are explained in Sect. 8.2. In Sect. 8.3, the estimated results on panel data analysis are shown. Sect. 8.4 presents conclusion.

8.2 Estimated Model and Data Set

8.2.1 The Market Value Model: Brynjolfsson and Yang Type Model

In this section, the authors explain the market value model as used by Brynjolfsson and Yang (1997). This model relates the assets of the firm to the market value of the firm.

After the leading research of Tobin (1969), the concept of Tobin's q is used when the relationship between firm market value and capital investment is observed. Hayashi and Inoue (1991) are famous for analysis of Tobin's q using the multiple capitals.

A firm decides the expenditure of investment vector and variable cost vector, $I(t) = (I_1(t), \ldots, I_J(t))$ and $N(t) = (N_1(t), \ldots, N_L(t))$, respectively, to maximize the integrated value $V(0)$ of the discounted present value of the profit $\pi(t)$ at time $t = 0$.[5] $\mu(t)$ is the discount function, and $F(\cdot, \cdot)$ is the production function. In addition, the organizational adjustment cost function $\Gamma(I(t), K(t))$, which is the additional cost by the investment, is assumed. A capital stock vector $K(t) = (K_1(t), \ldots, K_J(t))$ that includes fixed capital and information system asset at time t is regarded as satisfying the constraints of the sum total of the capital investment and the capital stock which deducts the depreciation vector δ.

$$V(0) = \max_{\{N(t), I(t)\}_{t=0}^{\infty}} \int_0^{\infty} \pi(t)\mu(t)dt \tag{8.1}$$

$$\text{subject to } \dot{K}_j(t) = I_j(t) - \delta_j K_j(t) \text{ for all } j = 1, \ldots, J \tag{8.2}$$

where

$$\pi(t) = F(K(t), N(t)) - \sum_{j=1}^{J} I_j(t) - \sum_{l=1}^{L} N_l(t). \tag{8.3}$$

$F(\cdot, \cdot)$ is a linear homogeneous function and continuously differentiable over K and N. $\Gamma(\cdot, \cdot)$ is a linear homogeneous function and continuously

---

[5]Inada condition is used for the production function. The theorem of Mangasarian is used to satisfy the sufficient condition of the optimization. See the mathematical appendix in appendix C for details.

differentiable over I and K. $\Gamma(\cdot, \cdot)$ is a increasing function of I and is convex and non-negative without fixed cost.

If the adjustment cost does not exist, buying the firm is equivalent to buying the assets of the firm separately. In short, the firm value equals the sum total of the asset value.

$$V(0) = \sum_{j=1}^{J} K_j(0) \tag{8.4}$$

If the adjustment cost exists, the firm value may exceed the sum total of the value of separate assets. Brynjolfsson and Yang regard the excess firm value as intangible assets formed, when assets are integrated as a firm. In this case, the firm value becomes the sum total of the assets weighted by the shadow price.

$$V(0) = \sum_{j=1}^{J} \lambda_j K_j(0) \tag{8.5}$$

It is possible to calculate the organization investment $(\lambda_j - 1)K_j(0)$ in comparison with the market value of the assets which are sold separately and assets which become part of the firm.

The immeasurable intangible assets except for adjustment cost again are defined as $(\nu_j - 1)K_j(0)$. If the optimization problem of Eq. 8.3 from Eq. 8.1 are rearranged using this condition, the market value equation becomes as follows.

$$V(0) = \sum_{j=1}^{J} \lambda_j(0)(\nu_j(0) - 1)K_j(0) + \sum_{j=1}^{J} \lambda_j K_j(0)$$
$$= \sum_{j=1}^{J} \nu_j(0)\lambda_j(0)K_j(0) \tag{8.6}$$

In short, the parameters of this estimated model $\nu_j(0)\lambda_j(0)$ are larger than 1 because of the adjustment cost and intangible fixed assets.

When three types of capital are assumed, the authors obtain Eq. 8.7 as the estimate equation.

$$V_i(t) = \alpha_i + \beta_{IS}K_{IS,i} + \beta_c K_{c,i} + \beta_o K_{o,i} + \varepsilon_i(t) \tag{8.7}$$

where i and t are indices of firms and time. $K_{IS,i}$, $K_{c,i}$ and $K_{o,i}$ represent information system assets, current assets, and other assets. $\varepsilon_i(t)$ represents the error term. The estimated parameters are α_i, β_{IS}, β_c and β_o. The authors can regard Eq. 8.7 as hedonic regression which estimates the shadow price of the assets using cross-section and time series data.[6]

In this chapter, the authors use the outsourcing variable (commitment ratio to related subsidiaries and outside companies occupied for the system development cost) as the proxy variable of intangible assets. The authors analyze whether outsourcing raises the market value of the firm when outsourcing advances with IT.

8.2.2 Data Sources and Construction

Data used in this analysis are from the questionnaire data on information system assets of 1995, 1996, and 1998 and financial statements. The data of information system investment on the questionnaire survey include computer software investment and development personnel expenses. The questionnaire survey defines "information system investment" as an expenditure including the following items:

1. The installation cost of terminal equipment including mainframes, workstations, personal computers, cash dispensing (CD) machines, automatic teller machines (ATMs), and so on
2. Purchase fees and charges for computer software
3. Personnel expenses related to 1 and 2 above

By combining these items in the analysis of this chapter, two kinds of information system investment are considered:

1. Information System Investment I: computer hardware, computer software, peripheral equipment costs and personnel expenses of normal staff member in system development
2. Information system investment II: computer hardware, computer software, peripheral equipment costs

The authors cannot disregard the personnel expenses of information system development in the development cost of information systems. Although personnel expenses are included in the amount in the case of outsourcing of system development, personnel expenses in information system section are treated in the accounting of the personnel expenses of other employees.

Although personnel expenses are dealt with as a cost, it is necessary to treat them as investment, if it is to be seen as human resource. The analysis of this chapter uses the value converted to assets. The authors did not use the

---

[6]The hedonic approach is to express the quality as a total of individual functions and properties using an objective index with the concept that the quality of the product is liberated into function and properties. Equation 8.3 regards the market value as a total of the assets; it is similar to this approach. See Shiratsuka (1998) for the hedonic approach to Japanese banking.

data of banks that did not clearly answer about the amount of information system investment in the questionnaire.[7]

Banking information systems are composed of multiple systems.[8] The systems, such as accounting systems, narrowly-defined information systems, securities operation and management system, and so on, are developed and operated at different times. Our questionnaire asked for the development and operation commencement times for the system, implying that of accounting systems. Similarly, the system development cost in the questionnaire implied that for the development of the accounting system.

However, it was possible to check the consistency of past years of the system of banks that answered in the first and second questionnaires because the history of the system including the third-generation on-line system was a topic of the third questionnaire. Many banks that replied in the first questionnaire also replied in the second questionnaire. On the basis of their data, it is possible to grasp the information system investment value from 1995 to 1996.

Nationwide banks tend to renew the systems completely except after periods of system failure. On the contrary, many regional banks conduct part renewal of the systems at regular intervals until complete renewal. The investment cost of part renewal of the systems may be included in the operating cost.

Two banks that answered the questionnaires used joint development centers. In these cases, the investment amounts of each bank are not clear. The authors used the total amount of the joint development cost, and divided by the number of participating banks as the contributing amount for each bank.

Some banks supplied labor input data as the number of employees rather than as man months. This required adjustment of the data following telephone interviews to ensure that all data were consistent and comparable.

Some banks did not provide the number of information system development workers or the average salary. The data of these banks were excluded from the estimate of Information System Investment I.

Thus, the number of the data is different according to the definition of the information system investment. Data for this analysis were over 3 years, from 33 banks, with 58 unbalanced panel data. 5 nationwide banks were included, and regional banks were thoroughly reported from all area of Japan. 18 banks replied over 2 years.

One question used in the first and second questionnaires asked for the development cost of current systems excluding the costs of labor and sub-

---

[7]In the USA, a database on installed computer equipment is maintained by Computer Intelligence Infocorp Co. Brynjolfsson group used all capital stocks of the computer such as center arithmetic unit, personal computer, peripheral equipment. This data does not contain information processing and communication equipment. The computer equipment with which information system staff were not concerned was not considered.

[8]See Fig. 1.4.

contracting.[9] In the third questionnaire, this question was changed so that it asked for the total amount of system development costs including rental since the third-generation on-line system was established. Fundamental analysis of the results uses only the first and second questionnaires for this particular question, because the question in the third questionnaire is not comparable with the others.

The method for obtaining stock value of the information system from the total cost of the information system of the questionnaire is explained here. The permanent inventory method is used for the stock variable. Equation 8.8 is the discrete form of Eq. 8.2. The authors assume that the depreciation period of the computer hardware is 5 years and the depreciation period of the computer software is 5 years. Because the questionnaire survey contains both computer hardware and computer software, the depreciation period of 5 years was adopted. Therefore, the depreciation ratio is assumed to be 0.37.

$$K_j(t) = I_j(t) + (1 - \delta)K_j(t - 1) \tag{8.8}$$

Depreciation ratio, δ, and system investment increasing rate, g, are assumed to be constant. The authors obtained system investment increasing rate from the amount of increase in information system investment of each bank.

Expanding Eq. 8.8 on the assumption of depreciation ratio, δ, and system investment increasing rate, g, and carrying out calculation by geometric series gives the approximate equation of the relation between capital stock at time t and capital stock at time $t - 1$ as follows:

$$K_j(t - 1) \approx \frac{I_j(t)}{(g + \delta)} \tag{8.9}$$

Next, book value of capital stock is obtained from Eqs. 8.8 and 8.9. The investment value in every year is obtained by dividing total sum of information system adjusted at growth rate, g, at the years from operation start point of time.

Nishikawa (2001) indicated the importance of computer price deflation for international comparison. He indicated that deflation of computers and associated equipment of the USA Bureau of Economic Analysis (BEA) is decreasing faster than the deflation of computers according to the Japanese wholesale price index by Statistical Survey Department of the Bank of Japan.

It is not appropriate to use computer hardware deflation with extreme changes, because the computer software is also included for the data used in this chapter. The authors use Japanese wholesale price index including computer hardware by Statistical Survey Department of the Bank of Japan.

---

[9]Motohashi (2003) used the outsourcing of production as the variable expressed collaborative activities with other firms. The variable was not statistically significant.

The book value of the computer assets is converted to real value divided by the domestic wholesale price index which traces back to the past years of the system. Multiplying the real value by the domestic wholesale price index at each point of time leads to reacquisition cost of the system assets.[10]

The computer hardware assets must be excluded from movable assets, when movable asset is used as an explanatory variable. The reacquisition cost was calculated by the same procedure as the reacquisition cost of information system assets, after the office machinery was subtracted from the amount of movable assets. This is because computer hardware and computer software in the questionnaire were not clearly divided.

Loan and bills discounted are used as liquid assets, because loan and bills discounted account for a significant proportion of banking assets. However, the authors use the value subtracted the bad debt from the loan and bills discounted, because bad debt is included in the loan and bills discounted. By this means, the loan and bills discounted are assumed to be evaluating the market price.

Multiplying the issued number of shares by the stock price and adding the book value of debt gives the total market value. Details of the number of shares, movable asset, office machineries, debt, loan and bills discounted, and bad debt are obtained from financial statements for each bank. The authors used monthly average stock prices of each bank at the end of the fiscal year as the stock price data.[11]

The outsourcing variables for banks undertaking outsourcing to outside firms were made from the questionnaire data. The authors calculated two types of ratios as followed:

$$OUT_1 = \frac{OSE}{ISDC + OSE} \tag{8.10}$$

and

$$OUT_2 = \frac{OSE + ACSE}{ISDC + OSE + ACSE}, \tag{8.11}$$

where OSE, $ISDC$, and $ACSE$ are outside subcontract expenses, information system development cost with personnel expenses, and affiliated company subcontract expenses, representatively.

---

[10]Brynjolfsson and Yang (1997), and Brynjolfsson, Hitt and Yang (2000, 2002) used the total sum after 4 years of depreciation including the fiscal year concerned as fixed capital stock in this term.

[11]The monthly average stock prices were obtained from Toyo Keizai Inc. stock price CD-ROM 1999.

8.3 Estimated Results

In this section, the authors analyze how the computer assets influence the stock market value of the banking industry.[12] Because the sizes of the banks that replied in the questionnaire were diverse, it is impossible to analyze the data without normalizing the bank characteristics. As a method for controlling the effects of unchangeable firm characteristics on the time, the authors can use the fixed effects model. Unchangeable individual effects on the time in the fixed effects model are estimated only by the number of banks. However, the degrees of freedom may drastically decrease from the fixed effects model, because the data number for the estimate in this chapter is limited.

In the meantime, it is possible to prevent the decrease in the degrees of freedom in case of the variable effects model to control the firm characteristics using the probability distribution. Although it is possible to reject one of two models by the Hausman test, the authors compare the fixed effects model with the variable effects model in this chapter.

The introduction of the trend variable is also examined to control the time-related factor. To begin with, the authors show unbalanced panel analysis using the questionnaire data from 1995 and 1996. Although banks that replied in only one of these years are also included in the questionnaire data, the authors estimated the model using unbalanced panel data in which the number of replied years of each bank was not even to ensure the number of data. Equation 8.12 shows the correlation matrix of the data for the analysis. K_c, K_{IS_I}, $K_{IS_{II}}$ and K_o are loan and bills discounted, Information System Asset I, information system asset II, and movable asset minus office machineries.

$$\begin{array}{c} \\ K_c \\ K_{IS_I} \\ K_{IS_{II}} \\ K_o \end{array} \begin{array}{cccc} K_c & K_{IS_I} & K_{IS_{II}} & K_o \\ \begin{pmatrix} 1.0000 & 0.8197 & 0.8308 & 0.9833 \\ 0.8197 & 1.0000 & 0.9999 & 0.8592 \\ 0.8308 & 0.9999 & 1.0000 & 0.8665 \\ 0.9833 & 0.8592 & 0.8665 & 1.0000 \end{pmatrix} \end{array} \qquad (8.12)$$

The data number is 42 for 27 banks. The authors calculated variance inflation factor (VIF) standard on this correlation coefficient.[13] The value between loan and bills discounted and current assets is 30.174 and is larger than 10. Equation 8.12 shows multicollinearity between the two variables. Therefore, the variable that the proportion occupied in the assets was small

---

[12]We apply TSP 4.5 (TSP International, Stanford, CA) and the PANEL procedures were used in the analysis.

[13]Variance inflation factor (VIF) is expressed as followed:

$$VIF_{ij} = \frac{1}{1 - R_{ij}^2}$$

where $R_i^2 j$ is the coefficient of determination when the i-th explanatory variable is regressed to the j-th explanatory variable.

(movable asset minus office machinery) should be excluded from the estimated formula.

Tables 8.1 and 8.2 show the estimated results for unbalanced panel data. The estimates of the fixed effects model are within estimates and the one of the random effects model are variance components estimates.

Table 8.1. Estimated result using unbalanced panel data (Information System I)

Parameter	Within estimates	Variance components estimates
$\hat{\alpha}$		$-3960.94 \ (-2.42499)^{**}$
$\hat{\beta}_{IS_I}$	$12.1760 \ (1.90301)^{*}$	$17.8402 \ (2.43034)^{**}$
$\hat{\beta}_c$	$1.70649 \ (40.6406)^{***}$	$1.72533 \ (103.622)^{***}$
Adj.R^2	0.999959	0.997904

27 banks, 42 data

F statistic: $F(24, 19) = 34.013, p = 0.00$

χ^2 statistic: $\chi^2(2) = 5.3117, p = 0.0702$

$^{*}p < 0.10, \ ^{**}p < 0.05, \ ^{***}p < 0.01$

Table 8.2. Estimated result using unbalanced panel data (Information System II)

Parameter	Within estimates	Variance components estimates
$\hat{\alpha}$		$-3948.33 \ (-2.42473)^{**}$
$\hat{\beta}_{IS_{II}}$	$13.9230 \ (1.98546)^{*}$	$20.0032 \ (2.44443)^{***}$
$\hat{\beta}_c$	$1.65462 \ (40.5188)^{***}$	$1.72284 \ (103.228)^{***}$
Adj.R^2	0.999960	0.997917

27 banks, 42 data

F statistic: $F(24, 19) = 33.853, p = 0.00$

χ^2 statistic: $\chi^2(2) = 4.7572, p = 0.0927$

$^{*}p < 0.10, \ ^{**}p < 0.05, \ ^{***}p < 0.01$

In the fixed effects model, the null hypothesis that the individual effect becomes a similar value was rejected for both of Information System Investment I and II. In short, the firm characteristics appear in individual effects of the fixed effects model. Because the correlation between loan and bills discounted and firm value is large, all coefficients of determination and coefficients of determination adjusted for the degrees of freedom exceed 0.99 in all estimates.

When there is correlation between the individual effect and the explanatory variable, the estimate of the fixed effects model is consistent and efficient.

The estimate of the random effects model is neither consistent nor efficient. Conversely, when there is no correlation between the individual effect and the explanatory variable, the estimate of the random effects model is consistent and efficient. The estimate of the fixed effects model is neither consistent nor efficient. In short, when there is correlation between the individual effect and the explanatory variable, the fixed effects model will be adopted. When there is no correlation between the explanatory variable and the individual effect, the random effects model will be adopted.

The Hausman test is a test of null hypothesis (H_0): that random effects would be consistent and efficient, versus the alternative hypothesis (H_1): that random effects would be inconsistent. The explanatory variable may correlate with the firm specific factor (individual effect), when the fixed effects model is adopted.

In the analysis of the unbalanced panel, the p-value of χ^2 statistic of the null hypothesis that there is no correlation between the explanatory variable and the individual effect is 0.0702 in the case of Information System Assets I and 0.0927 in the case of Information System Asset I. In short, the null hypothesis in both of information system assets I and Information System Assets II are rejected at the 10% significance level. The fixed effects models are adapted. However, it is not possible to reject the null hypothesis of being the random effects model, when the 5% significance level is adopted. Therefore, the estimated results of the fixed effects model and the random effects model are compared.

For this estimate, data from banks that answered only one questionnaire are also included, because the unbalanced panel data are used. The t-value of the parameter of information system assets shows that Information System Assets II works at a 1% significance level in the random effects model. In the case of the fixed effects model, it becomes significant at the 10% level.

Analysis of the Information System Assets I works at a 5% significance level in the variable effects model. In the case of the fixed effects model, it becomes significant at the 10% significance level.

The result shows that a rise of $1 in Information System Assets II leads to a increase in stock market value of about $20 in the case of the random effects model and about $14 in the case of the fixed effects model. The loan and bills discounted variable works significantly even in all estimated results at the 1% significance level.

The regression coefficient of the information system assets in the random effects model is bigger than that of the fixed effects model. The regression coefficient of information System Assets I with human resources is smaller than that of Information System Assets II without human resources. The fixed effects model was adopted when the level of significance was 10%. The difference of characteristics between regional banks and nationwide banks probably leads to large value of the significance level. The estimated result of the unchangeable individual effect on time correlated with system assets variable will be affected, if the time-related firm specific characteristics are not skillfully controlled. There is no guarantee which parameters of regional banks are equal to those of nationwide banks, even if the trend variable is introduced into the model to remove the time-related factor of the bank characteristics. The number of nationwide banks for the analysis is far less than that of regional banks.

Then, the authors analyze using the balance panel data of regional banks which removed nationwide banks. The data number is decreased in this case and it becomes 30 data of 15 banks. Tables 8.3 and 8.4 show the analytical result.

Table 8.3. Estimates of regional banks using balanced panel data (Information System I)

Parameter	Within estimates	Variance components estimates
$\hat{\alpha}$		-451.171 (-0.479228)
$\hat{\beta}_{IS_I}$	15.5819 (1.44287)	16.4630 (1.90154)*
$\hat{\beta}_c$	1.49592 (3.21724)***	1.45987 (20.6974)***
Adj.R^2	0.999811	0.979470

15 banks, 30 data

F statistic: $F(14, 13) = 33.963, p = 0.000$

χ^2 statistic: $\chi^2(2) = 24.271, p = 0.000$

$^*p < 0.10, ^{**}p < 0.05, ^{***}p < 0.01$

Table 8.4. Estimates of regional banks using balanced panel data (Information System II)

Parameter	Within estimates	Variance components estimates
$\hat{\alpha}$		-568.271 (-0.611730)
$\hat{\beta}_{IS_{II}}$	10.8521 (1.976624)*	13.6826 (1.64628)
$\hat{\beta}_c$	1.65462 (40.5188)***	1.72284 (103.228)***
Adj.R^2	0.999960	0.979701

15 banks, 30 data

F statistic: $F(14, 13) = 30.662, p = 0.000$

χ^2 statistic: $\chi^2(2) = 22.709, p = 0.000$

$^*p < 0.10, ^{**}p < 0.05, ^{***}p < 0.01$

Thus, the p-value of the Hausman test (χ^2 statistic) decreases drastically when the analysis is carried out by limiting it to regional banks and the probability that the fixed effects model is adopted rises. In comparison with the estimated result of Tables 8.1 and 8.2 with nationwide banks included, the regression coefficient of information system assets without nationwide banks is smaller than that with nationwide banks. This shows the possibility in which the market value effect of information system of regional banks is smaller than that of nationwide banks. The possibility that firm specific characteristics which change over time affects the value of the fixed individual effect is high when the authors estimate by pooling a small number of nationwide banks with many regional banks. The authors cannot estimate the market value model of nationwide banks, because there is a restriction in the number of questionnaire data.

Finally, the authors analyze the existence of intangible assets that are related to information system assets using questionnaire data. IT leads to an increase of human capital and organization reform, and, as a result, raises the firm value. The organization reform would lead to flat organizational structure and also to outsourcing. The authors introduce the ratio of subcontractor expenses for the system development cost into the estimated model as a proxy variable of outsourcing, because the outsourcing seems to advance the information system. It is possible to make several kind indexes whether the authors add human resources of in-house system development personnel to the system development cost and whether the authors add subcontract expenses of subsidiary to the outsourcing. The estimate was carried out using 38 data of 24 banks in which the preparation of outsourcing variable (OUT_i, for $i = 1, 2$ in Eqs. 8.10 and 8.11) was possible within the data used in Tables 8.1 and 8.2.

Tables 8.5 and 8.6 show the result using OUT_2 in Eq. 8.11 as the outsourcing assets variable which express the organization. The outsourcing variable is not significant even in all estimates. Long-term analysis is necessary to see

Table 8.5. Unbalanced panel estimates with the outsourcing variable (Information System I)

Parameter	Within estimates	Variance components estimates
$\hat{\alpha}$		-4415.26 $(-2.3996)^{**}$
$\hat{\beta}_{IS_I}$	11.2526 $(1.2576)^{*}$	18.3066 $(2.5881)^{***}$
$\hat{\beta}_c$	1.81768 $(5.7662)^{***}$	1.72536 $(103.96)^{***}$
Adj.R^2	0.99996	0.99781

24 banks, 38 data

F statistic: $F(23, 11) = 71.455, p = 0.000$

χ^2 statistic: $\chi^2(2) = 8.249, p = 0.0411$

$^{*}p < 0.10, ^{**}p < 0.05, ^{***}p < 0.01$

Table 8.6. Unbalanced panel estimates with the outsourcing variable (Information System II)

Parameter	Within estimates	Variance components estimates
$\hat{\alpha}$		$-4382.13 \; (-2.3924)^{**}$
$\hat{\beta}_{IS_{II}}$	$12.6845 \; (2.3683)^{***}$	$20.3809 \; (2.8117)^{***}$
$\hat{\beta}_c$	$1.77295 \; (5.6477)^{***}$	$1.72296 \; (103.49)^{***}$
Adj.R^2	0.99996	0.99680

24 banks, 38 data

F statistic: $F(23, 11) = 73.275, p = 0.000$

χ^2 statistic: $\chi^2(2) = 7.5674, p = 0.0559$

$^{**}p < 0.05, \; ^{***}p < 0.01$

whether the progress of IT brings about the subcontracting.[14] Although only the outsourcing of information system was taken up, the progress of IT probably affects the organization structure of the other departments in advance.

8.4 Conclusion

The results of the analysis of this chapter show that the market value effect of information system assets was bigger than those of other assets. The increase in the stock market value by the an increase in information system assets of $1 far exceeded the increase in the market value generated by other assets. The estimated result of Information System Assets I with human resources of information system development gave a small regression coefficient of information system assets.

The estimated result of the market value model demonstrates that the possibility of adopting fixed effects model is high. However, the possibility of rejecting the variable effects model becomes lower when nationwide banks are included. This may be due to the accuracy of the questionnaire data. In the questionnaire, a number of banks replied that it was difficult to clearly separate the development and operation costs. Regional banks appear to deal with the expenditure of partial renewal of the information system as an operating cost. Therefore, there may be differences of accuracy in the questionnaire data between nationwide banks and regional banks. In addition, it may be impossible to control firm characteristics by only individual effects of the fixed effects model that the scale differential among banks is constant with time. The pos-

---

[14]Motohashi (2003) used the outsourcing of production as the variable expressed collaborative activities with other firms, the variable was not statistically significant in the estimate of the Cobb-Douglas production function.

sibility in which the distribution of the error term has not normalized also cannot be denied.

The estimated result without the inclusion of nationwide banks shows a higher possibility of adopting the fixed effects model than with the inclusion of nationwide banks. It is necessary to use the data of more many banks to stabilize the estimated result when including nationwide banks. Although these cautionary notes exist, the high possibility of adopting the fixed effects model implies the correlation between firm characteristics and explanatory variables. Individual effects may indicate the correlation of intangible assets with information system assets. It is necessary to conduct a questionnaire on intangible assets such as organization reform and human capital in the banking industry to analyze the individual effect.

Analysis of Information System Investment Using Public Data

T. Takemura, S. Watanabe, and Y. Ukai

9.1 Introduction

In Chap. 8, we examined the propositions of Brynjolfsson and Yang (1997) and analyzed information system investment models in the banking industry that included relations between adjustment costs and unmeasured correlated intangible assets. We found the estimated value of Tobin's q far exceeded 1 point for information system asset by using questionnaire data.[1] By conducting three questionnaires, data from a total of 42 banks was accumulated, although 27 of the banks answered the questionnaire more than once. About 30% of the banks listed on the stock exchanges in Japan are represented in the questionnaire data, and thus we cannot deny the possibility that bias may arise. Under the theoretical framework of Brynjolfsson and Yang (1997), we analyze panel data made on the basis of public data concerning financial statements of banking industry in this chapter.

Since the latter half of the 1980s, much literature relating to information technology investment has emerged from the use of data at both industry level and firm level, as was examined in Chaps. 3 and 4. In many studies, information technology capitals were defined as installed computer equipment. This means that computer software assets yet to be installed are not included as

---

[1]Tobin's q is estimated in the following ways: one method uses the assumption "the marginal q related to investment in equipment is equal to the average q," to estimate the average q and examine its relation with investment in equipment. The second method is to estimate a production function and directly pursue the marginal q from marginal productivity of capital. This chapter adopts the former method and estimates the average q. Refer to Hayashi (1982), and Hayashi and Inoue (1991) for further information. Refer to Ban (1989) for a discussion on the validity of a model estimating Tobin's q.

part of the information technology capitals.[2] The reasons for this are as follows: data sets relating to only computer software asset are not available, and it is too difficult to distinguish computer software asset from computer equipment. Actually, by 2001 there existed no empirical study of productivity about computer software investment not only at the firm level, but also at the macro and industry level. Ukai and Takemura (2001) quickly paid attention to the significance of software assets in the banking industry, and estimated Tobin's q for computer software assets per employer. However, they did not estimate Tobin's q including computer equipment. We are interested in exploring the effects of information system investment including not only computer software investment, but also that in computer equipment. Therefore, we study the effect of information system investment in this chapter as the sum of computer software assets and computer equipment. We also study the effects of two components; computer software and computer equipment, separately.

This chapter is organized as follows: Sect. 9.2 considers the econometric model (for panel data) including some controlled variables, and gives explanation for data sets we have constructed; and Sect. 9.3 shows some results of our analysis by using both balanced and unbalanced panel data. Finally, Sect. 9.4 gives conclusion.

9.2 Analytical Tools and Data Set

In Chap. 8, we sketched the banking industry's stock market value model (Brynjolfsson and Yang Type model) on the basis of the dynamic optimization problem. The basic structure of the model is referred as the "Tobin's q" literature and describes the relationship between the valuation of capital goods and the market value of each firm. Refer to Subsect. 8.2.1 and Appendix C for the theoretical framework. Under this framework, we estimate Tobin's q.

In this chapter, we use the banking industry's stock market value model per employer. Here, even if we use Brynjolfsson and Yang model per employee, the essence of our analysis cannot be lost by assumptions on both a production function and organizational adjustment cost.[3]

---

[2]See Brynjolfsson and Yang (1997), and Brynjolfsson, Hitt, and Yang (2000, 2002).

[3]In Chap. 8, we assumed that a production function and organizational adjustment cost function were homogeneous functions of degree 1 over capital stocks, expenditures in variable costs and capital investments (constant return to scale), and twice differentiable. We also assumed that the organizational adjustment cost function was increasing and convex in investment, with no fixed cost and is nonnegative everywhere.

The number of employees in each bank was almost constant in the 1993-1999 period (in the period of the post-third-generation on-line system), although this consideration does not include the hiring temporary employees. One of the aims of

9.2.1 Formation: Model for Panel Data

We incorporate some new control variables into our econometric model, to account for sample heterogeneity, and analyze our model by using the panel data. To empirically estimate the relationship in Eq. 8.6 as the result of dynamic optimal behavior, we specify the panel data model as follows:

$$V_i(t) = \alpha_i + \sum_{j \in J} \beta_j K_{j,i}(t) + \sum_{m \in M} \gamma_m X_{m,i}(t) + \varepsilon_i(t) \tag{9.1}$$

where i, t, j, and m are indices of banks, period, different capital goods, and control variables. $V_i(t)$, $K_{j,i}$, and $X_{m,i}$ represent the maximized market value of each bank, the stock of capital assets, and control variables. α_i, β_j, γ_m and $\varepsilon_i(t)$ are the individual effect term, the market shadow price for the stock of capital assets (considered unmeasured correlated intangible assets), effect of control variables, and an error term.

We test two kinds of hypothesis for the panel data model. One tests the hypothesis that the constant terms are all equal in the F tests.[4] Another tests whether the individual effects are not correlated with the other regressors, as is assumed in the random effects model. Hauseman's specification test is used to test for orthogonality of the random effects and the regressors. Refer to Baltagi (1995), Greene (1997), and Mátyás and Sevestre (1992) for details of these tests for the hypothesis of the panel data model.

9.2.2 Data Sources and Construction

In this chapter, we construct a panel data set on the basis of banks' accounting information used in financial statements. To estimate the parameters in Eq. 9.1, we divide assets into two categories: information system and the other balance sheet assets. For other control variables, we use the 1998 R&D standards, X_A, and the operation period of the on-line system in each bank, X_O.

Most numerical information collected from the disclosed financial statements was of book value. Therefore, all asset values need to be converted into current value with the use of an appropriate price index. Unfortunately, we cannot obtain an appropriate price index for some of the assets, and so

---

the introduction of computer systems in banks was to reduce the number of employee, as described in Chap. 1. This aim has been realized over the long term, but did not change much in the short term. In Chap. 8 we collected data of employees engaged in system operation at branch offices from questionnaires. On the other hand, financial statements also disclose such data. Therefore, we control data in this chapter by converting data into "data per employer."

[4]The t ratio for the constant term can be used for a test of the hypothesis that it equals zero. This hypothesis is for one specific group, but is typically not useful for testing in this regression context.

each asset is used as the book value in this chapter. The explanation of data processing for analysis is shown in the this section.[5]

Total Market Value

We use the average annual stock price as the stock price. It is this stock price that we use to obtain the simple average of the maximum stock prices, $P_{\max}$, and minimum stock prices, $P_{\min}$, of each bank at the end of the fiscal year. We gain stock market value by multiplying the stock price by the total number of issued shares, TI. Furthermore, by adding debt, D, into the stock market value, we gain the total market value. The total market value divided by the number of employees, EMP, is

$$V = \frac{\frac{1}{2}(P_{\max} + P_{\min})TI + D}{EMP}.$$

Computer Software Assets

As described in Chap. 6, accounting procedure of computer software is arbitrary, because there was no clear accounting standard for computer software in financial statements until the 1998 fiscal year. Therefore, some banks have charged computer software as an asset, while others have charged it as a cost. In this study, banks that charged computer software as a cost are not included. The computer software asset, $SOFT$, divided by the number of employees is

$$K_S = \frac{SOFT}{EMP}.$$

Computer Equipment

Office machinery is classified in financial statements as a tangible fixed asset. After the 1990s, some banks began renting and/or leasing not only personal computers and peripheral equipment, and CD/ATM-related instruments, but also mainframes and host computers. "Situation of equipment" in financial statements includes information on computer equipment leased as specified by the First Subcommittee of Business Accounting Council (1993) after the 1993 fiscal year. In particular, we pay attention to the computer equipment by financial lease transaction.[6] We obtain the stock variable by accumulating the computer equipment for lease after 1993. Here we add the computer equipment

---

[5]The authors thank Prof. K. Suda, who is a leader in the field of positive accounting, for advice.

[6]Naturally, when we consider leased equipment, lease charges include interest rates. However, to simplify analysis, we assume that the interest rate is zero. In addition, our data does not include computer equipment used through both operational lease and rental.

by financial lease transaction, EQ, and office machinery, OM, to give the total computer equipment. Then, the total computer equipment divided by the number of employees gives

$$K_H = \frac{EQ + OM}{EMP}.$$

Information System Assets

As described in Sect. 5.2, we defined some information system investments. Analyses in this chapter used the concept of "Information System Asset II." By this definition, we know that Information System Asset II does not include organization change and human resource. Thus, we regard the information system assets per employee, K_{INF}, as that obtained by adding the computer software assets per employee, K_S, and computer equipment per employee, K_H. Refer to Chap. 6 for the concepts on these financial statements. That is,

$$K_{INF} = K_S + K_H.$$

Loan and Bills Discounted

Loan and bills discounted represent one of assets that are distinct from information system assets. However, loan and bills discounted on financial statements include bad debt.[7] In particular, bad debt has become a social problem since the collapse of the bubble economy. Fig. 9.1 shows that the share of bad debt in the loan and bills discounted has changed drastically after 1993.

In 1993, bad debt was obligated to disclose at first. At that time, the share of bad debt in loan and bills discounted was less than 2.0% for most banks. However, the reporting of bad debt was partly modified in 1998, causing a change of distribution. In particular, bad debt of most banks hold more than 2.0% in loan and bills discounted. Here, as the same as in Chap. 8, we subtract bad debt, BD, from loan and bills discounted, LB.[8] When this loan and bills discounted is divided by the number of employees, EMP, is,

$$K_B = \frac{LB - BD}{EMP}.$$

----

[7]Although we recognize that it is an important social problem, we do not especially refer to the problem of bad debt. We allow this problem to be discussed elsewhere and consider only the classification of bad debt. Before 1997, bad debt was defined as the amount made up of failed future credits, credits in arrears, interest rate reductions of tax credits, and credits for management support companies. After 1998, bad debt was defined as the amount made up by failed future credits in a business year, credits in arrears, credits in arrears by more than 3 months, and terms and conditions of loan relief credit.

[8]When we consider the bad debt in each bank that increases rapidly, the question of whether we can control assets by processing of this data remains. For further detail, refer to Chap. 8.

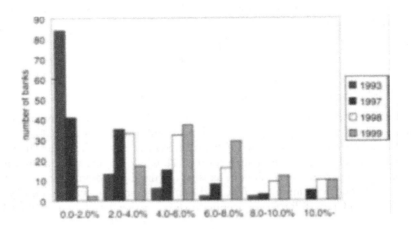

Fig. 9.1. Share of bad debt in loan and bills discounted in Japanese banks for 1993 and the 1997-1999 period

1998 R&D Standards

"The Statement of opinions on the establishment of accounting standards for research and development costs" was implemented in March 1998 by the Business Accounting Council, Ministry of Finance, as we see Chap. 6. This statement includes "accounting standard for research and development costs." (the authors call it "1998 R&D standards" hereafter) Preparation of the accounting standard did not advance until the 1998 R&D Standards were announced and implemented. This fact may imply that in some firms, including banks, placed low priority on information systems, especially computer software, despite of playing an important role in management.[9]

The number of banks that wrote computer software in their financial statements as an asset increased because the 1998 R&D standards for computer software was implemented in 1998. This accounting standard requires legal disclosure and therefore influences the behavior of firms. Therefore, we assume that a given accounting standard effects a change in the total market value, and we introduce the dummy variable, X_A, for the 1998 R&D standards. This variable is assigned as 0 for cases before this accounting standard was implemented and is assigned as 1 for cases after this standard was implemented, concretely 1999 fiscal year. That is,

$$X_A = \begin{cases} 1 & \text{if the 1998 R\&D standards is implemented} \\ 0 & \text{otherwise} \end{cases}.$$

---

[9]See Sect. 6.6.

Operation Period of On-line System

Fig. 9.2 shows the histogram of timing for introduction of the current on-line system at 1999. The latest on-line system includes a accounting system, narrowly-defined information system, security system, and so on[10].

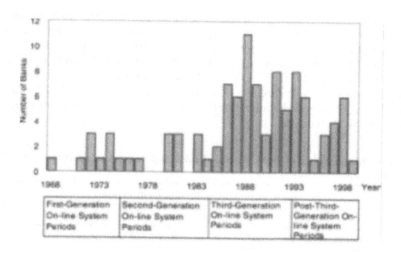

Fig. 9.2. Number of Japanese banks using on-line systems at the time

We collected and processed these data based on information provided by the banks under "history of firm" in financial statements. Unfortunately, some banks do not provide this information, and we summarized it for a total of 97 banks. Most on-line systems of banks in 1999 were described as the "third-generation on-line system" or the "new total on-line system." Some banks operated with a "second-generation on-line system," while others quickly operated a system called "fourth-generation on-line system." Refer to Chap. 1 for a discussion of the progress of on-line systems.

Operation period of on-line system represents transit time from introduction to 1999. Dummy variable for operation period of the on-line system, X_O, is assigned a value of 0 if the operation period exceeds 5 years. Otherwise, it is assigned. That is,

$$X_O = \begin{cases} 1 & \text{if the operation period exceeds 5 years} \\ 0 & \text{otherwise} \end{cases}.$$

---

[10]See Fig. 1.4 and narrowly-defined information system are described in Chap. 5.

We assume the standard of period for the on-line system is 5 years. This is based on depreciation period of 5 or 6 years, and about 5 years has passed since the post-third-generation on-line system period began.

9.2.3 Unbalanced Panel Data and Balanced Panel Data

We made a panel data set on the basis of data in the 1993-1999 period. The number of Japanese banks in this period had not been constant; one reason is that some banks failed, merged, or newly listed on the market. Another reason is that most banks did not write the computer software assets in their financial statements. Refer to Chap. 6 for a detailed discussion of how banks include computer software assets in financial statements. Table 9.1 shows the number of years and the number of banks that write computer software assets in financial statements.

Table 9.1. Number of bank disclosed information concerning computer software assets in financial statements

	Duration (years)						
	1	2	3	4	5	6	7
Number of banks 34(28)		3	9	3	6	7	12

The number of banks in parentheses shows the number of banks that write computer software assets in financial statements only in 1999

As shown in Table 9.1, the number of banks shows considerable variation with the number of years. Fortunately, even if there are some missing values in the data set, Eq. 9.1 can be estimated. We refer to such a data set as unbalanced panel data. On the other hand, if no missing values exist in the data set , then we refer to such a data set as balanced panel data.

If we analyze the data without regard for some banks failure and merger of banks, we may generate a survival bias. In other words, it is difficult to precisely analyze the banking industry in Japan if we only pay attention to the existing banks. Thus, by using an unbalanced panel data set, we can analyze the banking industry including failure or merger of banks.

We are interested in the banks that continue to write computer software assets in their financial statements. Therefore, we also analyze these banks by using balanced panel data. This means that balanced panel data analysis assumes that banks voluntarily disclose the necessary information. As seen in Table 9.1, unbalanced panel data analysis covers 74 banks and balanced panel data analysis covers 12 banks.

9.3 Estimation

When multicolinearity occurs between explanatory variables, a statistical problem is generated. Thus, we have to check whether multicolinearity between explanatory variables occurs. In this chapter, we adopt the variance inflation factor (VIF) criterion as one of the measures for multicolinearity.[11] If the value of VIF exceeds 10, multicolinearity between explanatory variables exists.[12]

As a result of examining for multicolinearity, we have no multicolinearity between variables used in both balanced panel data analysis and unbalanced panel data analysis. The correlation matrix of unbalanced panel data, M_U, and balanced panel data, M_B, are as follows:

$$
M_U = \begin{array}{c} \\ K_{INV} \\ K_S \\ K_H \\ K_B \\ X_A \\ X_O \end{array}
\begin{array}{cccccc}
K_{INV} & K_S & K_H & K_B & X_A & X_O \\
\left(\begin{array}{cccccc}
1.00 & 0.79 & 0.85 & 0.61 & 0.11 & -0.10 \\
0.79 & 1.00 & 0.35 & 0.63 & 0.05 & 0.08 \\
0.85 & 0.35 & 1.00 & 0.38 & 0.12 & -0.22 \\
0.61 & 0.63 & 0.38 & 1.00 & 0.06 & -0.20 \\
0.11 & 0.05 & 0.12 & 0.06 & 1.00 & -0.26 \\
-0.10 & 0.08 & -0.22 & -0.20 & -0.26 & 1.00
\end{array} \right)
\end{array},
$$

$$
M_B = \begin{array}{c} \\ K_{INV} \\ K_S \\ K_H \\ K_B \\ X_A \\ X_O \end{array}
\begin{array}{cccccc}
K_{INV} & K_S & K_H & K_B & X_A & X_O \\
\left(\begin{array}{cccccc}
1.00 & 0.73 & 0.81 & 0.53 & 0.04 & -0.11 \\
0.73 & 1.00 & 0.18 & 0.50 & -0.01 & 0.17 \\
0.81 & 0.18 & 1.00 & 0.33 & 0.06 & -0.31 \\
0.53 & 0.50 & 0.33 & 1.00 & 0.37 & -0.41 \\
0.04 & -0.01 & 0.06 & 0.37 & 1.00 & -0.28 \\
-0.11 & 0.17 & -0.31 & -0.41 & -0.28 & 1.00
\end{array} \right)
\end{array}.
$$

Note that for simplicity, we put information system assets, computer software assets, and computer equipment into the same correlation matrix.

Furthermore, the correlation matrix of unbalanced panel data, M_{U1}, for nationwide banks and the correlation matrix of balanced panel data, M_{U2}, for regional banks are as follows:

$$
M_{U1} = \begin{array}{c} \\ K_{INV} \\ K_S \\ K_H \\ K_B \\ X_A \\ X_O \end{array}
\begin{array}{cccccc}
K_{INV} & K_S & K_H & K_B & X_A & X_O \\
\left(\begin{array}{cccccc}
1.00 & 0.61 & 0.82 & 0.49 & 0.37 & -0.24 \\
0.61 & 1.00 & 0.06 & 0.59 & 0.51 & -0.05 \\
0.82 & 0.06 & 1.00 & 0.20 & 0.09 & -0.26 \\
0.49 & 0.59 & 0.20 & 1.00 & 0.37 & -0.47 \\
0.37 & 0.51 & 0.09 & 0.37 & 1.00 & -0.09 \\
-0.24 & -0.05 & -0.26 & -0.47 & -0.09 & 1.00
\end{array} \right)
\end{array},
$$

---

[11] Refer to Maddala (1992) for further discussion of the VIF criterion.
[12] See Sect. 8.3 for expression.

$$M_{U2} = \begin{array}{c} \\ K_{INV} \\ K_S \\ K_H \\ K_B \\ X_A \\ X_O \end{array} \begin{pmatrix} K_{INV} & K_S & K_H & K_B & X_A & X_O \\ 1.00 & 0.74 & 0.84 & 0.39 & 0.11 & -0.04 \\ 0.74 & 1.00 & 0.27 & 0.39 & 0.01 & 0.17 \\ 0.84 & 0.27 & 1.00 & 0.25 & 0.15 & -0.20 \\ 0.39 & 0.39 & 0.25 & 1.00 & 0.07 & -0.17 \\ 0.11 & 0.01 & 0.15 & 0.07 & 1.00 & -0.29 \\ -0.04 & 0.17 & -0.20 & -0.17 & -0.29 & 1.00 \end{pmatrix}.$$

We apply TSP 4.5 (TSP International, Stanford, CA) to our panel data analysis for statistical processing.

9.3.1 Estimation (1) Unbalanced Panel Data Analysis

Table 9.2 shows the results of the analysis by using information system assets, K_{INF}, and loan and bills discounted, K_B, as explanatory variables, and the 1998 R&D standards, X_A, and the operation period of the on-line system, X_O, as control variables.

Table 9.2. Estimated result using unbalanced panel data I

Parameter	Within estimates	Variance components estimates
$\hat{\alpha}$		-150.541^{***} (-3.281)
$\hat{\beta}_{INF}$	11.229^* (1.883)	8.495 (1.478)
$\hat{\beta}_B$	1.276^{***} (16.465)	1.775^{***} (44.546)
$\hat{\gamma}_A$	44.876^{**} (2.321)	-3.267 (-0.190)
$\hat{\gamma}_O$	14.758 (0.744)	28.613 (1.514)
Adj.R^2	0.993	0.946

74 banks, 234 data

F statistic: $F(73, 153) = 22.153, p = 0.00$

χ^2 statistic: $\chi^2(4) = 59.923, p = 0.00$

$^*p < 0.10,$ $^{**}p < 0.05,$ $^{***}p < 0.01$

These results are derived from unbalanced panel data of 74 banks over 7 years; in particular, we have 234 data sets because of missing values. The F value for the simultaneous significance test of individual effect is 22.153 and the critical value of F is 9.587. This implies there is individual effect at a 1% level of significance. Furthermore, in checking the Hausman test, a null hypothesis is rejected at a 1% level of significance and the fixed effects model is accepted because χ^2 value of four degrees of freedom is 59.923 and the critical value of χ^2 is 13.277.

In the fixed effects model (see within estimates column in Table 9.2), an estimated coefficient of information system assets, $\hat{\beta}_{INF}$, becomes about

11.2 points and far exceeds 1 point. An estimated coefficient of loan and bills discounted, $\hat{\beta}_B$, also exceeds 1 point and becomes about 1.3 points. For control variables, an estimated coefficient of the 1998 R&D standards, $\hat{\gamma}_A$, becomes about 44.9 points and positive, but an estimated coefficient of the operation period of the on-line system, $\hat{\gamma}_O$, is not statistically significant.

Table 9.3 shows the results of the analysis by using computer software assets, K_S, computer equipment, K_H, and loan and bills discounted, K_B, as explanatory variables, and X_A and X_O as control variables.

Table 9.3. Estimated result using unbalanced panel data II

Parameter	Within estimates	Variance components estimates
$\hat{\alpha}$		-119.421^{***} (-2.583)
$\hat{\beta}_S$	31.550^{**} (2.333)	52.470^{***} (4.227)
$\hat{\beta}_H$	2.543 (0.323)	-10.785 (-1.442)
$\hat{\beta}_B$	1.290^{***} (16.642)	1.744^{***} (43.218)
$\hat{\gamma}_A$	40.946^{**} (2.114)	-4.698 (-0.275)
$\hat{\gamma}_O$	5.327 (0.260)	7.768 (0.398)
Adj.R^2	0.993	0.947

74 banks, 234 data

F statistic: $F(73, 155) = 21.958, p = 0.00$

χ^2 statistic: $\chi^2(5) = 52.464, p = 0.00$

$^{**}p < 0.05,$ $^{***}p < 0.01$

In this case, we divide information system assets into computer software assets and computer equipment. This analysis consists of unbalanced panel data of 74 banks over 7 years; in particular, there are 234 data sets as in Table 9.2.

The F value for the simultaneous significance test of individual effect is 21.958 and the critical value of F is 9.521. This implies that individual effects are significant at the 1% level. Furthermore, in checking the Hausman test, a null hypothesis is rejected at a 1% level of significance and a fixed effects model is accepted because the value of five degrees of freedom is 52.464 and the critical value of is 15.086.

In the fixed effects model (see within estimates column in Table 9.3), the estimated coefficient of computer software assets, $\hat{\beta}_S$, becomes about 31.6 points and far exceeds 1 point. The estimated coefficient of loan and bills discounted, $\hat{\beta}_B$, also exceeds 1 point and becomes about 1.3 points. For control variables, the estimated coefficient of the accounting standard of the 1998 R&D standards, $\hat{\gamma}_A$, becomes about 40.9 points and positive. However, neither the estimated coefficient of computer equipment, $\hat{\beta}_H$, nor the estimated coefficient for the operation period of the on-line system, $\hat{\gamma}_O$, are statistically significant. That is, we know that both $\hat{\beta}_H$ and $\hat{\gamma}_O$ are zero.

In each case of using unbalanced panel data, the fixed effects model is accepted. This implies that each bank has firm characteristic. Although it cannot be observed directly from bank data, it shows that each bank may possess own information management style or culture, which may include organizational culture, brand, technology, and management strategies.

One million dollars of information system investment (resp. computer software investment) increases the total market value by about 11.2 (resp. about 31.6) million dollars and one million dollars of loan and bills discounted increases the total market value by about 1.3 million dollars. On the contrary, increased investment in computer equipment does not have an effect on the total market value.

In Tables 9.2 and 9.3, we analyzed unbalanced panel data without distinguishing nationwide banks from regional banks. However, there are some discrepancies between the information system strategies of nationwide banks and those of regional banks, as described in Chap. 2. These discrepancies may cause different effects of information system investment between nationwide banks and regional banks. Thus, we separately analyze unbalanced panel data for nationwide banks and regional banks.

Nationwide Banks

Table 9.4 shows the result of unbalanced panel data analysis using the data of nationwide banks. Then we use the information system assets, K_{INF}, and loan and bills discounted, K_B, as explanatory variables, and the 1998 R&D standards, X_A, and the operation period of the on-line system, X_O, as control variables. Note that this analysis consists of unbalanced panel data of 8 banks over 7 years, and in particular we have only 36 data sets because of defalcation values.

The F value for the simultaneous significance test of individual effect is 3.200. On the contrary, the critical value of F is 3.454. Therefore, it is difficult to conclude that an individual effect exists. Moreover, in checking the Hausman test, a null hypothesis is not rejected and a random effect model is accepted because the χ^2 value of four degrees of freedom is 11.970 and the critical value of χ^2 is 13.277.

In the random effect model (see variance components estimates in Table 9.4), the estimated coefficients of the information system assets, $\hat{\beta}_{INF}$, and the 1998 R&D standards, $\hat{\gamma}_A$, are not statistically significant. On the other hand, statistically significant coefficients of loan and bills discounted, $\hat{\beta}_B$, and the operation period of the on-line system, $\hat{\gamma}_O$, are about 1.5 points and 473.5 points, respectively. The statistically significant individual effect, $\hat{\alpha}$, is −836.659, which means that the individual effect is negative in this case.

Table 9.4. Estimated result using unbalanced panel data III (nationwide banks)

Parameter	Within estimates	Variance components estimates
$\hat{\alpha}$		-836.659^{***} (-4.130)
$\hat{\beta}_{INF}$	1.730 (0.058)	-15.434 (-0.600)
$\hat{\beta}_B$	1.500^{***} (5.972)	2.147^{***} (23.119)
$\hat{\gamma}_A$	-1.561 (-0.019)	-58.907 (-0.753)
$\hat{\gamma}_O$	123.070 (0.640)	473.465^{***} (4.290)
Adj.R^2	0.976	0.963

8 banks, 36 data

F statistic: $F(7, 24) = 3.200, p = 0.07$

χ^2 statistic: $\chi^2(4) = 11.970, p = 0.17$

$^{***}p < 0.01$

Furthermore, Table 9.5 shows the analysis results for nationwide banks when the information system asset, K_{INF}, is divided into computer software assets, K_S, and computer equipment, K_H.

Table 9.5. Estimated result using unbalanced panel data IV (nationwide banks)

Parameter	Within estimates	Variance components estimates
$\hat{\alpha}$		-805.451^{***} (-3.968)
$\hat{\beta}_S$	30.858 (0.528)	52.341 (1.020)
$\hat{\beta}_H$	-9.938 (-0.273)	-33.656 (-1.172)
$\hat{\beta}_B$	1.521^{***} (5.913)	2.084^{***} (20.312)
$\hat{\gamma}_A$	-13.336 (-0.155)	-92.025 (-1.125)
$\hat{\gamma}_O$	150.936 (0.752)	458.171^{***} (4.175)
Adj.R^2	0.976	0.964

8 banks, 36 data

F statistic: $F(7, 23) = 2.868, p = 0.06$

χ^2 statistic: $\chi^2(5) = 10.091, p = 0.07$

$^{***}p < 0.01$

The F value for the simultaneous significance test of individual effect is 2.868. On the contrary, the critical value of F is 3.310. Thus, it is difficult to conclude that an individual effect exists as in the above case. Moreover, in checking the Hausman test, a null hypothesis is not rejected and a random effect model is accepted because the χ^2 value of five degrees of freedom is 10.091 and the critical value of χ^2 is 15.086.

In the random effect model (see variance components estimates in Table 9.5), the estimated coefficients of computer software assets, $\hat{\beta}_S$, computer equipment, $\hat{\beta}_H$, and the 1998 R&D standards, $\hat{\gamma}_A$, were not statistically significant. On the other hand, statistically significant coefficients of loan and bills discounted, $\hat{\beta}_B$, and the operation period of the on-line system, $\hat{\gamma}_O$, were about 2.1 points and 458.2 points, respectively. The statistically significant individual effect, $\hat{\alpha}$, is -805.5 and is negative in this case.

Regional Banks

Table 9.6 shows the results of the analysis of unbalanced panel data using regional bank data. We used information system assets, K_{INF}, and loan and bills discounted, K_B, as explanatory variables, and the 1998 R&D standards, X_A, and the operation period of the on-line system, X_O, as control variables. Note that this analysis consists of unbalanced panel data of 66 banks over 7 years, and in particular we have 198 data sets.

Table 9.6. Estimated result using unbalanced panel data V (regional banks)

Parameter	Within estimates	Variance components estimates
$\hat{\alpha}$		62.567 (1.292)
$\hat{\beta}_{INF}$	11.368*** (2.894)	13.558*** (3.524)
$\hat{\beta}_B$	1.089*** (16.089)	1.479*** (28.719)
$\hat{\gamma}_A$	72.378*** (5.221)	28.657** (2.270)
$\hat{\gamma}_O$	13.955 (1.149)	15.944 (1.340)
Adj.R^2	0.992	40.851

66 banks, 198 data

F statistic: $F(65, 128) = 53.279, p = 0.00$

χ^2 statistic: $\chi^2(4) = 81.288, p = 0.00$

**$p < 0.05$, ***$p < 0.01$

The F value for the simultaneous significance test of individual effect is 53.279 and the critical value of F is 9.206. This implies that an individual effect exists at a 1% level of significance. Furthermore, in checking the Hausman test, a null hypothesis is rejected at a 1% level of significance and a fixed effects model is accepted because the χ^2 value of four degrees of freedom is 81.288 and the critical value of χ^2 is 13.277.

In the fixed effects model (see within estimates in Table 9.6), the estimated coefficient of information system assets, $\hat{\beta}_{INF}$, becomes about 11.4 points and far exceeds 1 point. The estimated coefficient of loan and bills discounted, $\hat{\beta}_B$, also exceeds 1 point and becomes about 1.1 points. For control variables, the estimated coefficient of the 1998 R&D standards, $\hat{\gamma}_A$, becomes about 72.4 points and positive, but the estimated coefficient of the operation period of the on-line system, $\hat{\gamma}_O$, is not statistically significant.

Table 9.7 shows the result for regional banks when the information system asset, K_{INF}, is divided into computer software assets, K_S, and computer equipment, K_H.

Table 9.7. Estimated result using unbalanced panel data VI (regional banks)

Parameter	Within estimates	Variance components estimates
$\hat{\alpha}$		67.337 (1.405)
$\hat{\beta}_S$	32.334*** (3.398)	50.148*** (5.584)
$\hat{\beta}_H$	2.688 (0.510)	2.335 (−0.452)
$\hat{\beta}_B$	1.115*** (16.563)	1.482*** (29.135)
$\hat{\gamma}_A$	67.800*** (4.934)	25.808** (2.075)
$\hat{\gamma}_O$	3.259 (0.256)	−2.731 (−0.220)
Adj.R^2	0.992	0.852

66 banks, 198 data

F statistic: $F(65, 127) = 54.478, p = 0.00$

χ^2 statistic: $\chi^2(5) = 74.513, p = 0.00$

$^{**}p < 0.05$, $^{***}p < 0.01$

The F value for the simultaneous significance test of individual effect is 54.478 and the critical value of F is 9.134. This implies that an individual effect exists at a 1% significance level. Furthermore, in checking the Hausman test, a null hypothesis is rejected at a 1% level of significance and a fixed effects model is accepted because the χ^2 value for five degrees of freedom is 74.513 and the critical value of χ^2 is 15.086.

In the fixed effects model (see within estimates in Table 9.7), the estimated coefficient of computer software assets, $\hat{\beta}_S$, becomes about 32.3 points and far exceeds 1 point. The estimated coefficient of loan and bills discounted, $\hat{\beta}_B$, also exceeds 1 point and becomes about 1.1 points. For control variables, the estimated coefficient of the 1998 R&D standards, $\hat{\gamma}_A$, becomes about 67.8 points and positive. However, both the estimated coefficient of computer equipment, $\hat{\beta}_H$, and the estimated coefficient of operation period of the on-line system, $\hat{\gamma}_O$, are not statistically significant.

In regional banks, one million dollars of information system investment (resp. computer software investment) increases the total market value by about 11.4 (resp. about 32.3) million dollars and one million dollars of loan and bills discounted increases the total market value by about 1.1 million dollars. On the contrary, increased investment in computer equipment does not have an effect on the total market value in both nationwide banks and regional banks.

In Tables 9.4 and 9.5, we show the results of unbalanced panel data analyses for nationwide banks, while in Tables 9.6 and 9.7, we show the results of the same analyses for regional banks. The findings are summarized as follows: first of all, in unbalanced panel data analysis, the random effect model is accepted by panel data and the fixed effects model is accepted by regional bank data. This means that regional banks have firm characteristics, not nationwide banks.[13] Second, information system assets of nationwide banks do not impact on their total market values, but information system assets of regional banks give their total market values considerably high positive impact. Finally, the computer software assets of regional banks would give considerably higher total market values than the information system assets.

9.3.2 Estimation (2) Balanced Panel Data Analysis

In Subsect. 9.3.1, we used and analyzed unbalanced panel data that the banks disclose their computer software assets in their financial statements for more than one years. In this section, we focus on balanced panel data that the banks disclose their computer software in their financial statements for 7 years.[14]

Table 9.8 shows the results of a balanced panel data analysis. Here, we use information system assets, K_{INF}, and loan and bills discounted, K_B, as explanatory variables, and the 1998 R&D standards, X_A, and the operation period of the on-line system, X_O, as control variables. Note that we have 84 data sets because these consist of balanced panel data of 12 banks for 7 years.

The F value for the simultaneous significance test of individual effect is 33.704 and the critical value of F is 4.862. This implies that there is an individual effect at a 1% level of significance. Furthermore, in checking the Hausman test, a null hypothesis is rejected at a 1% level of significance and the fixed effects model is accepted because the χ^2 value of three degrees of freedom is 47.008 and the critical value of χ^2 is 11.345.

In the fixed effects model (see within estimates in Table 9.8), the estimated coefficient of information system assets, $\hat{\beta}_{INF}$, is not statistically significant. The estimated coefficient of loan and bills discounted, $\hat{\beta}_B$, becomes about 1.1 points. For control variables, the estimated coefficients of both the1998 R&D standards, $\hat{\gamma}_A$, and the operation period of the on-line system, $\hat{\gamma}_O$, become statistically significant, and are approximately 74.5 points and 42.6 points, respectively.

---

[13]An assertion made in Part I that there are different behaviors toward information system investment for nationwide banks and regional banks, is supported by these results.

[14]See Table 9.1.

Table 9.8. Estimated result using balanced panel data I

Parameter	Within estimates	Variance components estimates
$\hat{\alpha}$		96.001* (1.698)
$\hat{\beta}_{INF}$	6.218 (1.130)	2.080 (0.384)
$\hat{\beta}_B$	1.081*** (15.317)	1.407*** (27.705)
$\hat{\gamma}_A$	73.485*** (4.326)	32.860** (2.065)
$\hat{\gamma}_O$	42.571*** (2.649)	50.520*** (3.207)
Adj. R^2	0.996	0.977

12 banks, 84 data

F statistic: $F(11, 68) = 33.704, p = 0.00$

χ^2 statistic: $\chi^2(3) = 47.008, p = 0.00$

$^*p < 0.10,\ ^{**}p < 0.05,\ ^{***}p < 0.01$

Table 9.9 shows the results for the analysis by using computer software assets, K_S, computer equipment, K_H, and loan and bills discounted, K_B, as explanatory variables, and X_A and X_O as control variables.

Table 9.9. Estimated result using balanced panel data II

Parameter	Within estimates	Variance components estimates
$\hat{\alpha}$		120.669** (2.162)
$\hat{\beta}_S$	28.974*** (2.644)	29.710*** (2.751)
$\hat{\beta}_H$	−3.248 (−0.488)	−9.183 (−1.406)
$\hat{\beta}_B$	1.077*** (15.783)	1.383*** (27.668)
$\hat{\gamma}_A$	71.252*** (4.328)	32.438** (2.101)
$\hat{\gamma}_O$	30.142* (1.837)	34.515** (2.131)
Adj. R^2	0.996	0.977

12 banks, 84 data

F statistic: $F(11, 67) = 35.097, p = 0.00$

χ^2 statistic: $\chi^2(4) = 46.363, p = 0.00$

$^*p < 0.10,\ ^{**}p < 0.05,\ ^{***}p < 0.01$

The F value for the simultaneous significance test of individual effect is 35.097 and the critical value of F is 4.790. This implies that there is an individual effect at a 1% level of significance. Furthermore, in checking the Hausman test, a null hypothesis is rejected at a 1% level of significance and a fixed effects model is accepted because the χ^2 value of four degrees of freedom is 46.363 and the critical value of χ^2 is 13.277.

In the fixed effects model (see within estimates in Table 9.9), the estimated coefficient of computer software assets, $\hat{\beta}_S$, becomes about 29 points and far exceeds 1 point. The estimated coefficient of loan and bills discounted, $\hat{\beta}_B$, also exceeds 1 points and becomes about 1.1 points. For control variables, the estimated coefficients of both the 1998 R&D standards, $\hat{\gamma}_A$, and the operation period of the on-line system, $\hat{\gamma}_O$, become statistically significant, and are approximately 71.3 points and 30.1 points, respectively. However, the estimated coefficient of computer equipment, $\hat{\beta}_H$, is not statistically significant.

In both Tables 9.8 and 9.9, the fixed effects model is accepted and we find that each bank that disclosing spontaneously has firm characteristic. This is the same result as that of the unbalanced panel data analysis. Moreover, one million dollars of computer software investment increases the total market value by about 29 million dollars and one million dollars of loan and bills discounted increases the total market value by about 1.1 million dollars. On the contrary, increased investment in computer equipment does not have an effect on the total market value.

Second, while the operation period dummy of the on-line system is not statistically significant in each unbalanced panel data analysis, the dummy variable is statistically significant in each balanced panel data analysis. That is, we find that updating or renewal of the information system gives positive effects to the total market value for banks that spontaneously disclose their information system assets.[15] Following the conclusions drawn in Chap. 1 concerning the efficiency of organizations, rationalization of office work, and so on, the operation period of an on-line system is reflected in the total market value of the bank.

Finally, we found that the 1998 R&D standards exerts a positive effect on total market value for each bank that discloses its information system assets. For banks that disclose spontaneously, implementing the 1998 R&D standards lets them recognize the importance of information system assets, especially computer software assets. In other words, if banks recognize information systems or computer software as management resources and disclose them in the financial statements, the actions would be reflected in the total market value.

9.4 Concluding Remarks: Future Directions and Opportunities for Research

In this chapter, we investigated and estimated Tobin's q for information system assets in the Japanese banking industry. At first, we found that the estimated value of Tobin's q for loan and bills discounted that deducted bad debt

---

[15]Here we suppose that there are few intervals in which the information system is updated if the system works for less than 6 years.

becomes 1.1-1.5 points, and the estimated value of Tobin's q for information system assets becomes 9.0-9.5 and far exceeds 1 point. Furthermore, when we divide the information system assets into computer software assets and computer equipment, the estimated value of Tobin's q for the former becomes about 30 points, but the estimated value for the latter becomes not statistically significant. This means that computer equipment does not affect the total market value. We can suggest the following reasons to explain this fact.[16] First of all, prices of computer equipment are not reflected in the total market value of existing equipment in use. In addition, the fall in prices of computer equipment does not reflect the total market value of the rapid improvement of information technology.

Second, the importance of the estimated values of information system assets and computer software assets becomes apparent. The difference may be observed by including computer equipment in information system assets. In other words, the effect of information system assets would be underestimated if the value of computer equipment, which does not affect the total market value, is included. From the theory of Tobin's q, the value of Tobin's q for information system assets still far exceeds 1 point.[17]

Third, the authors did not find effects of computer equipment nor computer software assets to the total market value of nationwide banks. Meanwhile, the authors found effect of computer software assets to the total market value of regional banks. This finding is appalling to our research. However, the authors expect the coming new generation on-line system brings about a positive effect on the total market value.[18]

Until the 1990s, interest decision theory and credit structure theory were the mainstream analytical tools for banks and the banking industry. These theories, however, shift toward finance theory; theory of portfolio selection, or capital market theory. The theory of Tobin's q that we investigate in this chapter is categorized as finance theory. Thus, from the viewpoint of portfolio selection, our analysis provides material for management use. In the field of accounting and business, it would be mainstream that administrators or researchers use financial indexes that are measured by return on investment (ROI), sales growth rate or sales income, and so on as indicators to management[19]. While these financial indexes are effective in short-term management,

---

[16] Generally, we do not assume that the used market exists in theory of Tobin's q.

[17] The fact that Tobin's q exceeds 1 point suggests that not only does the utility value of the firm's information system assets exceed the price, but also the stock price overestimates the firm's value. It is difficult to judge which possibility is realistic.

[18] Takemura (2003) analyzed the effects of computer soft asset in the Japanese banking industry. See Chap. 4. Takemura found the effects of computer software after introduction of the on-line system, but the effect disappeared in the latter half of the 1990s.

[19] Prasad and Harker (1997) used ROA (return on asset) and ROE (return on equity) as output of banking, and failed to find the positive productivity on IT capital in the USA banking industry.

they may not always be effective in medium-term or long-term management. We use data on financial statements to analyze banks, but there may be limitations in the way that banks are evaluated only by a management index on financial statements.[20]

Balanced score card (BSC) is a management assessment method attracting attention in the field of management accounting, and get the attention in Japan. The concept of BSC targets only activities that are directly connected with strategies and visions characteristics to firms. After considering financial and nonfinancial indexes, the manager plans strategies through setting and administration of the aims and indexes. Generally speaking, BSC consists of four perspectives: financial perspective, internal business perspective, customer perspective, and innovation and learning perspective; and the firms are evaluated by these perspectives. See Kaplan and Norton (1996) about details of BSC.[21]

After the analysis demonstrated in this chapter, we suggest that assessment of information system investment and/or the market value of banks should be measured with BSC.[22] Although we can evaluate bank management, we know that nonfinancial indexes are not always suited to financial statements. We can construct BSC in each industry or industrial branch by using the framework of BSC. Then, we regard the information system as one of the management indexes. We can analyze panel data including not only the quantity of the information system (financial index), but also the quality of the information system or type of organization. (nonfinancial index)[23] In a fact, Hitt and Brynjolfsson (1997), and Brynjolfsson, Hitt, and Yang (2000, 2002) analyzed information system investment with consideration to organization structure. Thus, we can analyze about internal business prospective by the same manner. Furthermore, we suppose that information system investments, for instance those related to CDs and ATMs, are reflected directly by customer satisfaction, and then analysis can be conducted from the customer's perspective. That is, we can link story of Chap. 2 with BSC.

---

[20]Some criticism has been made concerning the analysis of information systems using only the quantity.

[21]First of all, Kaplan and Norton (1996) describe that point of view and administration aims of BSC originally chosen at the firm's discretion. Although we can estimate the value of a firm, we suggest the problem which possibility comes to lack simultaneously. However, in addition to analysis at the firm level, expansion to the level of the industry is possible by establishing the method against new perspectives. Ogura and Shimazaki (2001) analyzed the Japanese financial industry by using the BSC approach.

[22]Of course, these perspectives are not independent and there exists complementarity among them. The BSC approach does not deny approach by existing financial statements.

[23]See Hitt and Brynjolfsson (1997) about relationships among information system investment, organization structure, and performance evaluation.

We may know that nonfinancial indexes are not always suited to financial statements. Therefore, we need to continue to collect questionnaire data for nonfinancial indexes. Constructing BSC implies that we can study the banking industry and information system investment in the banking industry.

10

Conclusion

Y. Ukai

The number of Japanese private banking institutions – nationwide banks, regional banks, second regional banks, credit associations, credit unions, long-term credit banks, agriculture cooperatives, and fisher cooperatives – decreased by 34.5% , and the number of those branches decreasing at a rate of 5% from March 1994 to March 1999. By contrast, the number of ATMs increased by 62% during the same period. Japanese citizens enjoy the largest number of ATMs per capita in the world.[1] As the authors described in Chap. 1, almost all of these ATMs had been connected with one another in the 1990s.

The Bank of Japan Financial Network System (Nichigin Net) started to operate its simultaneous settlement system linked with government bonds and cash in 1994, and also started to operate its simultaneous settlement system linked with corporate bonds and cash in 1998. The Domestic Funds Transfer System (Zengin System) started to operate its electronic data interchange system (EDI system) linked with commercial distribution data and financial data in 1996.[2] Tokyo Stock Exchange, Inc. and Osaka Securities Exchange Co., Ltd. moved all their transactions into trading systems that connected their exchange systems with the client machines of all members of the stock exchanges in 1999. At that time, about 30% of Japanese households had personal computers and there were estimated 17 million regular Internet users, which were more than 10% of the total number of worldwide Internet users.

Generally speaking, there was the widespread impression that information system investments in Japanese banks were allocated smaller budgets than in the United States and Western European banks, which resulted in inferior management. For instance, it was frequently reported in Japanese newspapers and magazines that even the top four Japanese mega banks each invested, at the most, only 60 million dollars annually in their information systems,

---

[1]Refer to the Center for Financial Industry Information Systems (1999), pp501-502.

[2]Refer to Subsects. 1.2.1-1.2.3 in Chap. 1.

although giant financial corporations in the United States annually invested more than 2 billion dollars each.

However, it would be very dangerous for us to determine the relative merit of investments simply on a budgetary scale, as is proved by the statistical analyses in Part III of this book. In Japanese society, where almost all economic transactions are implemented by cash, information systems concerning cash-handling services are the most sophisticated and effective in the world. Therefore, CD/ATM and ATM networks in Japan achieved the highest level of technology in the world. By contrast, the ATMs in the USA and Western Europe are extremely user-unfriendly by comparison, because transactions are restricted in small amount of cash and have small variety, and because machines frequently get out of order.

Countries such as the USA, Great Britain, Australia, and English-speaking Canada are traditionally check societies, and the banks in these areas have suffered from cost burdens of check processing. Their mailing and communication costs reach enormous levels from sending paper checks to their customers, reconfirmation of processing, and so on. This type of intercultural difference among transaction customs should deeply affect the information system investment in banking industry.

In addition, attention must be paid to the international differences in banking business. In the USA and Western Europe, where deregulation in financial markets started earlier than in Japan, almost all banks are conducting trust business, securities business, and insurance business in parallel with ordinary banking business. The disclosed investment value could be nominally overestimated, because the systemization costs of these various businesses were included in the total investment value of information technology in banks.

Along with the progress of deregulation in the Japanese financial market, information system investment values of entire financial corporations owned by holding companies should be compared with the values of banks in the USA and Europe. Moreover, the information system investment value of Japan Post, which was privatized in 2003 and has savings and insurance departments, should also be considered.

From the historical and institutional examination in Part I, the authors concluded that the information technology system in the Japanese banking industry was the optimal system for an economy mainly operated by cash transaction through its evolutionary process of the first, second, third, and post-third-generation on-line systems. The Chinese economy and the other Asian economies could learn a great deal from the Japanese information system because they also operate mainly by cash transactions.

In the theoretical examination in Part II, the authors concluded that the total factor productivity and the labor productivity as determined by macroeconomic models were not always suitable for analysis of information systems. On the other hand, econometric analyses of information systems by microeconomic models had several statistical problems, for example, data restriction, which should be solved in the near future. However, the microeconomic model would be more fruitful than the macroeconomic model by a variety of statistical analyses, in addition to the existing quantitative evaluation indexes, if the disclosure of firms' information proceeded gradually. The authors established the Research Center of Socionetwork Strategies from this standpoint.

Among their original statistical and institutional analyses of information systems in Part III, the authors first estimated 15 regression curves by the cross-section analysis in Chap. 7, with the dependent variable of system development investment in each bank, and with independent variables of lending outstanding, total assets, and total employees, based on the questionnaires from 1995 to 1998. Every regression curve had a positive increment, and was statistically significant. In addition, by using a dummy variable for management strategies related to the reaction to the central bank and supervisory authority, it was suggested that management strategy deeply affected the investment propensity of each bank.

In Chap. 8, the authors conducted panel data analyses by using individual bank data derived from questionnaires and interviews in order to estimate the increase in market value of each bank due to information system investment. As a result of this analysis, it was proved that an additional 1 dollar of information system assets has a positive effect of increasing the market value of each bank by 12-18 dollars, when computer hardware, software, and personnel costs of system development are summed as an information system asset.

In addition, the authors used questionnaire data to estimate the contribution to market value of each bank by the outsourcing of system development. However, this investigation failed to determine whether the outsourcing of system development boosted market value of each bank.

In Chap. 9, the authors conducted other panel data analyses by using individual bank data disclosed in financial statements in order to estimate the contribution of information system investment to the market value of each bank. This analysis proved that an additional 1 dollar of information system assets per employee has a positive effect of increasing the market value of each bank by 11 dollars, when summing up computer hardware and software as an information system asset. Moreover, it was proved that an additional 1 dollar of computer software assets per employee has a positive effect of increasing the market value of each bank by 32 dollars.

Also in Chap. 9, the authors conducted separate panel data analyses for computer hardware and computer software. This work reached the important conclusion that computer software assets have a statistically significant positive effect on the market value despite the insignificant effect of the hardware assets.

Finally, in comparing the questionnaire analyses of Chap. 8 with the financial statement analyses of Chap. 9, no difference between the two analyses for regional banks could be identified. On the other hand, a large analytical difference was observed for nationwide mega banks. Only continuous questionnaire surveys of nationwide mega banks might clarify whether this analytical difference is temporal or stems from an institutional deficiency concerning information disclosure in the banking industry.

The conclusions of analyses presented in this book object to the perception of the information technology revolution in today's financial industry. Almost all Japanese bankers have been insisting that the positive effect of computer hardware and software to market value in the banking industry is much greater than the effect in the manufacturing industry.[3] They are not correct. The real effect in the Japanese banking industry has no difference with the USA manufacturing industry in the golden age.

The conclusion that additional software investment is effective, but additional hardware investment is ineffective to a bank's market value should suggest a critical management strategy to the banking industry. The aggressive investment of software and related human resources should be an urgent consideration for the financial industry and the world economy in the twenty-first century.[4]

---

[3]Refer to Nishimura (2000). Mr. Masao Nishimura was CEO at the Industrial Bank of Japan in 2000. This article was one of the typical opinions that information technology had a definitely more important role in the financial industry than in the manufacturing industry.

[4]See Ukai (2002).

Part IV

Appendix

A

Information System Investment Questionnaires 1995-1997

1995 Information System Investment Questionnaire (Distributed Friday, February 17, 1995)

Workshop on Information System Investment
Professor Ukai Laboratory, Faculty of Informatics, Kansai University,
2-1-1 Ryozenji-cho, Takatsuki-shi, Osaka 569-1095, Japan
Tel: +81 (0)72 690 2452
Fax: +81 (0)72 690 2493

The purpose of this questionnaire is to obtain data about investment in all banking information systems including system related to strategy development, communication network with the outsiders, and office processing. Information system investment is defined as follows:

1. The installation cost of terminal equipment including mainframes, workstations, personal computers, cash dispenser (CD), automatic teller machines (ATMs), and so on
2. Purchase fees and charges for computer software
3. Personnel expenses related to 1 and 2 above.

When unable to answer a question or when the criterion upon which it is based does not apply, please write N/A. If there are relevant points that are not raised in the questionnaire, please inform the coordinating laboratory.

Question 1-1 When did your current information system begin operation?

Answer 1-1 (   )

Questions about Development
Question 1-2 What length of time was required from the start of system development to first operation of the system?
Answer 1-2 About (   ) years

Question 2-1 How much manpower (regular officials, keiretsu subsidiary officials, and other officials from outside firms) was needed for development of the existing system?
Answer 2-1 About (   ) persons

Question 2-2 How many regular staff members were engaged in the system development of your bank?
Answer 2-2 About (   ) persons

Question 2-3 How many keiretsu subsidiary officials were engaged in the system development of your bank?
Answer 2-3 About (   ) persons

Question 3-1 What was the approximate cost of development of the existing system? (Personnel expenses and outsourcing costs not considered.)
Answer 3-1 About (   ) hundred million yen

Question 3-2 What is the average age and the average annual income of regular staff members engaged in development of the existing system?
Answer 3-2 (   ) million yen

Question 3-3 What were the approximate outsourcing costs for the keiretsu subsidiary for development of the existing system?
Answer 3-3 About (   ) million yen

Question 3-4 What were the approximate outsourcing costs for other outside firms for development of the existing system?
Answer 3-4 About (   ) million yen

Questions about Operation
Question 4-1 How much manpower (regular officials, keiretsu subsidiary officials, and officials from other outside firms) per month is needed for operation of the existing system?
Answer 4-1 About (   ) persons per month

Question 4-2 How many regular staff members are engaged in system operation at your bank per month?
Answer 4-2 About (   ) persons per month

Question 4-3 How many keiretsu subsidiary officials are engaged in system operation at your bank per month?
Answer 4-3 About (   ) persons per month

Question 5-1 What are the annual operating expenses of the existing system including rental and/or lease fees and running costs? (Personnel expenses and outsourcing costs not considered.)
Answer 5-1 About (   ) million yen per year

Question 5-2 What is the average annual income of regular staff members engaged in operation of the existing system?
Answer 5-2 About (   ) million yen

Question 5-3 What are the outsourcing costs for the keiretsu subsidiary for operation of the existing system?
Answer 5-3 About (   ) million yen

Question 5-4 What are the outsourcing costs for other outside firms for operation of the existing system?
Answer 5-4 About (   ) million yen

Question 6 After what time is the existing system scheduled to be updated?
Answer 6 After about (   ) years

Question 7-1 When you update the current system, please rank the following items in order of your importance.
(   ) Technical obsolescence
(   ) Investment actions of other banks
(   ) New product developments
(   ) Administrative guidance of Bank of Japan and Ministry of Finance
(   ) Others

Question 7-2 If you answered "Others" in Question 7-1, please give specific investment criteria.

Question 8 If you have any comments regarding this questionnaire, please enter below.

Thank you for your cooperation. Please place the answer sheet in the envelope addressed to Kansai University, and post before the end of March.

Date		Respondent name	
Bank name		Job title	

1996 Information System Investment Questionnaire (Distributed Monday, February 19, 1996)

Workshop on Information System Investment
Professor Ukai Laboratory, Faculty of Informatics, Kansai University,
2-1-1 Ryozenji-cho, Takatsuki-shi, Osaka 569-1095, Japan
Tel: +81 (0)72 690 2452
Fax: +81 (0)72 690 2493

The purpose of this questionnaire is to obtain data about investment in all banking information systems including system related to strategy development, communication network with the outsiders, and office processing. Information system investment is defined as follows:

1. The installation cost of terminal equipment including mainframes, workstations, personal computers, cash dispenser (CD), automatic teller machines (ATMs), and so on
2. Purchase fees and charges for computer software
3. Personnel expenses related to 1 and 2 above.

When unable to answer a question or when the criterion upon which it is based does not apply, please write N/A. If there are relevant points that are not raised in the questionnaire, please inform the coordinating laboratory.

Question 1-1 When did your current information system begin operation?
Answer 1-1 (   )

Questions about Development
Question 1-2 What length of time was required from the start of system development to first operation of the system?
Answer 1-2 About (   ) years

Question 2-1 How much manpower (regular officials, keiretsu subsidiary officials, and other officials from outside firms) was needed for development of the existing system?
Answer 2-1 About (   ) persons

Question 2-2 How many regular staff members were engaged in the system development of your bank?
Answer 2-2 About (   ) persons

Question 2-3 How many keiretsu subsidiary officials were engaged in the system development of your bank?
Answer 2-3 About (   ) persons

Question 3-1 What was the approximate cost of development of the existing system? (Personnel expenses and outsourcing costs not considered.)
Answer 3-1 About (   ) hundred million yen

Question 3-2 What is the average age and the average annual income of regular staff members engaged in development of the existing system?
Answer 3-2-1 About (   ) years
Answer 3-2-2 Annual income about (   ) million yen

Question 3-3 What were the approximate outsourcing costs for the keiretsu subsidiary for development of the existing system?
Answer 3-3 About (   ) million yen

Question 3-4 What were the approximate outsourcing costs for other outside firms for development of the existing system?
Answer 3-4 About (   ) million yen

Questions about Operation
Question 4-1 How much manpower (regular officials, keiretsu subsidiary officials, and officials from other outside firms) per month is needed for operation of the existing system?
Answer 4-1 About (   ) persons per month

Question 4-2 How many regular staff members are engaged in system operation at your bank per month?
Answer 4-2 About (   ) persons per month

Question 4-3 How many keiretsu subsidiary officials are engaged in system operation at your bank per month?
Answer 4-3 About (   ) persons per month

Question 5-1 What are the annual operating expenses of the existing system, including rental and/or lease fees and running costs? (Personnel expenses and outsourcing costs not considered.)
Answer 5-1 About (   ) million yen per year

Question 5-2 What is the average age and the average annual income of regular staff members engaged in operation of the existing system?
Answer 5-2-1 About (   ) years
Answer 5-2-2 Annual income about (   ) million yen

Question 5-3 What are the outsourcing costs for the keiretsu subsidiary for operation of the existing system?
Answer 5-3 About (   ) million yen

Question 5-4 What are the outsourcing costs for other outside firms for operation of the existing system?
Answer 5-4 About (   ) million yen

Question 6 For the existing system, what kind of replacement policy is used? Please indicate total replacement (1) or partial replacement (2), and enter approximate time in parenthesis.

1. (   ) We plan total system replacement after (   ) years.

2. (   ) We plan partial system replacement every (   ) months.

3. (   ) We use (2) together with (1).

Question 7 When you update the current system, please rank the following items in order of your importance.
(   ) Technical obsolescence
(   ) Investment actions of other banks
(   ) New product developments
(   ) Administrative guidance of Bank of Japan and Ministry of Finance
(   ) Others (please specify)

Question 8 In the results of this questionnaire in 1995, a number of respondents indicated that it was difficult to differentiate between the costs of system development and the costs of system use. However, the argument was put forward that such a problem should not arise in modern business accounting. Please add your comments regarding this contention:

Thank you for your cooperation. Please place the answer sheet in the envelope addressed to Kansai University, and post before the end of March.

Date		Respondent name	
Bank name		Job title	
Telephone		Fax	
E-mail			

1997 Information System Investment Questionnaire (Distributed Monday, December 15, 1997)

Workshop on Information System Investment
Professor Ukai Laboratory, Faculty of Informatics, Kansai University,
2-1-1 Ryozenji-cho, Takatsuki-shi, Osaka 569-1095, Japan
Tel: +81 (0)72 690 2452
Fax: +81 (0)72 690 2493

The purpose of this questionnaire is to obtain data about investment in all banking information systems including system related to strategy development, communication network with the outsiders, and office processing. Information system investment is defined as follows:

1. The installation cost of terminal equipment including mainframes, workstations, personal computers, cash dispenser (CD), automatic teller machines (ATMs), and so on
2. Purchase fees and charges for computer software
3. Personnel expenses related to 1 and 2 above.

The completed questionnaire should be enclosed in an envelope addressed to the coordinating laboratory and should be posted before the end of December. If questions cannot be answered because of a lack of data, please leave the answer fields blank.

Questions about the Financial Deregulation of Japan

Question 1-1 In regard to preparation for the financial deregulation of Japan, please indicate whether a special team for information system investment was organized.
Answer 1-1 (   ) A team was organized (   ) A team was not organized

Question 1-2 If you selected "A team was not organized" in Question 1-1, please select one of the two responses below.
Answer 1-2 (   ) We will organize one in the future (   ) We do not plan to organize one

Question 1-3 If you selected "We do not plan to organize one" in Question 1-2, please select one of the two responses below.
Answer 1-3 (   ) We cope in an existing system sector (   ) There is no correspondence schedule

Question 1-4 If you selected "We cope in an existing system sector" in Question 1-3, please indicate the responses below (one or both) that apply.
Answer 1-4 (   ) It is coped with by manpower increase of an existing sector
(   ) It is coped with by budget increase of an existing sector

Questions about Information System
Question 2-1 When did your current accounting system begin operation?
Answer 2-1 (   )

Question 2-2 When did your current narrowly-defined information system begin operation?
Answer 2-2 (   )

Question 3-1 What length of time was required from the start of development of the current accounting system to first operation of the system?
Answer 3-1 About (   ) years

Question 3-2 What length of time was required from the start of development of the current narrowly-defined information system to first operation of the system?
Answer 3-2 About (   ) years

Question 3-3 Please indicate the timeline of the current main system development process except for the accounting and narrowly-defined information systems. Please describe the three main systems.
Answer 3-3-1 (   ) system
    Development start 19 (   )   Operation start 19 (   )
Answer 3-3-2 (   ) system
    Development start 19 (   )   Operation start 19 (   )
Answer 3-3-3 (   ) system
    Development start 19 (   )   Operation start 19 (   )

Questions about System Development Labor Force
Question 4-1 How much manpower (regular officials, keiretsu subsidiary officials, and officials from other outside firms) was required for development of the existing accounting system?
Answer 4-1 About (   ) man months (multiply no. of persons by months)

Question 4-2 How much regular staff manpower was engaged in the development of your accounting system?
Answer 4-2 About (   ) man months

Question 4-3 How much keiretsu subsidiary manpower was engaged in the development of your accounting system?
Answer 4-3 About (   ) man months

Question 4-4 How much manpower for accounting system development are assigned to each branch?
Answer 4-4 About (   ) man months

Question 5-1 How much manpower (regular officials, keiretsu subsidiary officials, and officials from other outside firms) was needed for development of the existing narrowly-defined information system?
Answer 5-1 About (   ) man months

Question 5-2 How much regular staff manpower are engaged in the development of the narrowly-defined information system of your bank?
Answer 5-2 About (   ) man months

Question 5-3 How much keiretsu subsidiary manpower are engaged in the development of the narrowly-defined information system of your bank?
Answer 5-3 About (   ) man months

Question 5-4 How much manpower for narrowly-defined information system development was assigned to each branch?
Answer 5-4 About (   ) man months

Question 6-1 How much manpower (regular officials, keiretsu subsidiary officials, and officials from other outside firms) is needed for development of the existing system nominated in Question 3-3-1?
Answer 6-1 About (   )man months

Question 6-2 How many regular staff members are engaged in the development of the system nominated in Question 3-3-1?
Answer 6-2 About (   ) man months

Question 6-3 How many keiretsu subsidiary officials are engaged in the development of the system nominated in Question 3-3-1?
Answer 6-3 About (   ) man months

Question 6-4 How many members for system development are assigned to each branch for the system nominated in Question 3-3-1?
Answer 6-4 About (   ) man months

Question 7-1 How much manpower (regular officials, keiretsu subsidiary officials, and officials from other outside firms) is needed for development of the existing system nominated in Question 3-3-2?
Answer 7-1 About (   ) persons

Question 7-2 How much regular staff manpower are engaged in the develop-

ment of the system nominated in Question 3-3-2?
Answer 7-2 About (   ) persons

Question 7-3 How many keiretsu subsidiary officials are engaged in the development of the system nominated in Question 3-3-2?
Answer 7-3 About (   ) persons

Question 7-4 How much manpower of system development are assigned to each branch for the system nominated in Question 3-3-2?
Answer 7-4 About (   ) man months

Question 8-1 How much manpower (regular officials, keiretsu subsidiary officials, and officials from other outside firms) is needed for development of the existing system nominated in Question 3-3-3?
Answer 8-1 About (   ) man months

Question 8-2 How much regular staff manpower are engaged in the development of the system nominated in Question 3-3-3?
Answer 8-2 About (   ) persons

Question 8-3 How much keiretsu subsidiary manpower are engaged in the development of the system nominated in Question 3-3-3?
Answer 8-3 About (   ) persons

Question 8-4 How much manpower of system development was assigned to each branch for the system nominated in Question 3-3-3?
Answer 8-4 About (   ) man months

Questions about System Development Costs
Question 9-1 What was the aggregate cost for system developments including rental charges after development of the third-generation on-line system? (Personnel expenses and outsourcing costs not considered.)
Answer 9-1 About (   ) hundred million yen

Question 9-2 What is the average age and the average annual income of regular staff members engaged in the system development after development of the third-generation on-line system?
Answer 9-2-1 About (   ) years
Answer 9-2-2 Annual income about (   ) million yen

Question 9-3 What are the outsourcing costs for the keiretsu subsidiary for system development after development of the third-generation on-line system?
Answer 9-3 About (   ) million yen

Question 9-4 What are the outsourcing costs for other outside firms for system

development after development of the third-generation on-line system?
Answer 9-4 About (   ) million yen

Questions about System Operation Labor Force
Question 10-1 How much manpower (regular officials, keiretsu subsidiary officials, and officials from other outside firms) is needed monthly for system operation after the development of the third-generation on-line system?
Answer 10-1 About (   ) persons

Question 10-2 How many regular staff members are engaged in the system operation after the development of the third-generation on-line system?
Answer 10-2 About (   ) persons

Question 10-3 How many keiretsu subsidiary officials are engaged in the system operation after the development of the third-generation on-line system?
Answer 10-3 About (   ) persons

Question 10-4 Is system operation included in official service regulations of generalist staff members in each branch?
Answer 10-4 (   ) It is included (   ) It is not included

Others
Question 11-1 What are the annual operating expenses of the system including rental and/or lease fees and running costs after the development of the third on-line system? (Personnel expenses and outsourcing costs not considered.)
Answer 11-1 About (   ) million yen per year

Question 11-2 What is the average age and the average annual income of regular staff members engaged in system operation?
Answer 11-2-1 About (   ) years
Answer 11-2-2 Annual income about (   ) million yen

Question 11-3 What are the outsourcing costs for the keiretsu subsidiary for the system operation?
Answer 11-3 About (   ) million yen

Question 11-4 What are the outsourcing costs for other outside firms for the system operation?
Answer 11-4 About (   ) million yen

Question 12-1 For the existing system, what kind of replacement policy is used? Please indicate policy type and enter time and budget estimates in parentheses.

1. (   ) We plan total system replacement after (   ) years, with a budget of (   ) million yen,
2. (   ) We plan partial system replacement every (   ) months, with a budget of (   ) million yen,
3. (   ) We use (2) together with (1), with a total budget of (   ) million yen

Question 12-2 In the budget mentioned above, how much is allocated to Y2K?
Answer 12-2 About (   ) million yen

Question 13 When you update the current system, please rank the following items in order of their importance.
(   ) Technical obsolescence
(   ) Investment actions of other banks
(   ) New product developments
(   ) Administrative guidance and suggestions of Bank of Japan and Ministry of Finance
(   ) Reduction of office workforce
(   ) Others (please specify)

By answering this questionnaire and entering your name, you can receive free notices of workshop meetings and interim reports. Please indicate below if you are interested in this offer.

Date		Respondent name	
Bank name		Job title	
Tel		Fax	
E-mail		Yes, please notify me ( )	
		No, do not notify me ( )	

B

Documents about Accounting Standards

Acknowledgement

This appendix is a copy of "Accounting Standards for Research and Development Costs," R-2-1, "FASF Japanese Accounting Standards on CD-ROM: Release 2002." The authors thank to Mr. Toshitaka Hagiwara, the Chiefs Director of Financial Accounting Standards Foundation, Tokyo, for his copyright permission.

Statement of Opinions on the Establishment of Accounting Standards for Research and Development Costs

March 13, 1998
Business Accounting Council

Establishment of Accounting Standards for Research and Development Costs

I. Background

The Business Accounting Council ("the Council") decided to consider accounting standards for research and development costs, regarded as important investment information; it started discussions in July 1997, and issued "Statement of Opinions on the Establishment or Accounting Standards for Research and Development Costs (Exposure Draft)" in December of the same year, to elicit opinions from various fields.

Referring to opinions received, the Council continued consideration and partially modified the contents of the Exposure Draft so as to now issue this "Statement of Opinions on the Establishment of Accounting Standards for Research and Development Costs."

II. Necessity of Establish Accounting Standards

Research and development is an important factor for the future profitability of any business. Recently, expenditures on research and development have become more considerable due to the shortening product cycle, the shorter period to catch up on new technologies, and the wider range and more advanced level of research and development. Thus, the importance of research and development has greatly increased in business activities. As a result, information as to the gross amount of research and development costs and the contents of research and development is regarded as important investment information, that casts light on businesses' managerial policies and profit forecasts.

Resembling research and development costs, in Japan there are the separate concepts of "research costs" and "development costs." The scope of these costs is not always clear, and whether or not to capitalize such costs as assets is optional for businesses; it has been pointed out that this vagueness obstructs comparability between Japanese and foreign businesses.

Under these circumstances, it is necessary to establish accounting standards for research and development costs, in order to elicit appropriate information on research and development, promote comparability among business entities, and contribute to the harmonization of standards internationally.

Along with the advance of computer technologies and the growth of the information-intensive society, the role of software has rapidly become more important in business activities, and expenditures for the production of software have increased. Software production processes include activities regarded as research and development, but there are no clear accounting standards for this endeavor, and arbitrary accounting methods are applied for such costs; therefore it is necessary to establish well-defined, uniform accounting standards.

These Accounting Standards are designed to define the scope of research and development included within software production processes, and also to establish accounting treatment for software production costs that are not regarded as research and development costs, with a view to establishing consistent accounting treatment for software production costs overall.

III. Key Points and Concepts

1. Definition of Research and Development
   The definition of research and development will, of course, help to determine the scope of research and development costs. To secure comparability between domestic and foreign businesses, in regard to research and development costs, the Standards are designed on the basis of the following definitions, taking into account definitions already adopted in other countries and the scope of activities already recognized, in practice, as research and development by businesses in Japan.

Research is "planned study and investigation aimed at discovering new knowledge", and development is "materialization of research findings, and other knowledge, in the plan or design of a new product, service or production method ("products"), or in the plan or design for a remarkable improvement in an existing product".

For example, even an improvement study carried out at a production site may satisfy the definition of "remarkable improvement" if it is clearly treated as a distinct project. However, quality control activities at a production site, and the management cost of dealing with complaints from customers are not included in research and development.

2. Charging Research and Development Costs to Expenses, as Incurred

It is necessary to achieve comparability among businesses on research and development costs, an important kind of investment information; therefore, it is not appropriate to leave businesses to decide arbitrarily whether to charge the costs to expenses, or to capitalize them as assets, as is allowed under the existing accounting method.

At the time when research and development costs are incurred, whether they will lead on to profit in the future is unforeseeable, and even if a research and development plan advances, and expectations for future profit grow, the realization of the future profit is still not assured. Therefore, the Council has decided that it is inappropriate to capitalize research and development costs as assets on the balance sheet.

If the Council were to require businesses to capitalize certain qualified costs as assets, the Council would have to define conditions for such capitalization. It is very difficult, however, to define any practical objectives and applicable criteria, and if abstract criteria were to be applicable in deciding whether or not to capitalize the costs as assets, comparability among business entities might be lost.

Therefore, the Standards require businesses to charge all research and development costs to expenses, as incurred.

3. Costs of Software Production

1) Since the relationship between costs of software production and future profit varies depending on the purpose of production, the accounting standards concerning costs of software production should be established to depend on the purpose of production, not on how the software is acquired. (e.g. whether produced in-house or purchased from outside)

Therefore, when software is targeted to serve a particular purpose, and is developed to completion by modifying software that has been purchased, or produced by a subcontractor, expenses for the purchasing or subcontracting are to be charged based on the purpose of production, as shown in 3), below.

2) The costs of producing software to be used for research and development are to be charged as research and development costs. Even if software is produced for any use other than research and development,

the portion of the production costs that corresponds to research and development is to be charged as research and development costs.

3) To establish accounting standards for the costs of software production, other than research and development costs, according to the purpose of production, the Council has decided to distinguish software for sale from software for internal use, and further to divide software for sale into custom-made software and software sold on the market.

(1) Custom-made software

Custom-made software is to be accounted for in the same way as contracted construction.

(2) Software for sale on the market

To sell software on the market, the manufacturer first produces the master version of the software (the duplicatable, completed product, hereafter referred to as the "master"), and then puts duplicates on the market.

Typically, in the production of the master, some processes correspond to research and development, and others to manufacturing; therefore, it is necessary to determine the finishing point of research and development, and so determine which costs should be treated as costs of software production, incurred after the finishing point.

(i) Finishing point of research and development

Research and development are processes to materialize new knowledge. In the course of creating software, therefore, research and development should mean all the activities undertaken until the completion of the master, to which the manufacturer attaches a product number, showing a clear intention to sell i.e. "the first merchandised master product."

This is because, when compared with research and development for manufactured products in general, the completion of the master is regarded as equivalent to completion of the design stage for a new mass-market product.

(ii) Treatment of costs of software production, after the completion of research and development

Expenses paid for production activities to improve or enhance functions of a master, or of purchased software, are to be included in assets, unless these activities are considered to bring about remarkable improvement.

(iii) However, expenses paid for function maintenance, such as debugging, are not to be included in costs of activities to improve or enhance functions, but charged to expenses, as incurred.

The Council has decided that the acquisition cost of a master is to be recorded as an intangible fixed asset. This is because the master itself is not sold, but used (copied) for production, like production equipment or machinery, and the master is protected by legal rights (copyright), and its acquisition cost is measured through proper cost accounting.

(3) Software for internal use

Costs to acquire, for internal use, software that is sure to yield profits or reduce costs in the future, are to be included in assets, and depreciated over the expected period of use, in tandem with the expected future profits.

The Council has decided to include the acquisition cost of the software in assets because, when any contract has been entered into to provide outsource services using software is very likely to yield profit, or cost savings.

However, when creating in-house custom-designed software for internal use, or subcontracting the creation of such software, the costs are to be charged to expenses, unless the software is very likely to yield profits, or cost savings.

(4) Acquisition costs of software included in intangible fixed assets are to be depreciated by a reasonable method, such as depreciation based on expected sales quantity, depending on the nature of the software. The amount of depreciation for any one fiscal year, however, is not to fall below the amount of depreciation that would be required each year for the rest of the depreciation period, if the amount of depreciation were the same every year.

In respect of the master of software that is for sale on the market, it may be reasonable to allocate depreciation based on expected sales profits.

In general, it may be reasonable to depreciate software for internal use by the straight-line method.

IV. Presentation

1. Presentation on Financial Statements

To achieve comparability between businesses as to expenditure on research and development, the total research and development costs included in general administrative expenses for the current period, and also the total included in production costs in the current period are to be disclosed in notes to their financial statements.

Research and development costs that are not included in production costs for the current period are to be reported under the account title of research and development costs, in general administrative expenses.

2. Description of Research and Development Activities

The format of descriptions on research and development activities (organization and results of research, etc.) should be standardized under "Status of Business" in the Annual Report, to facilitate comparison between entities.

If items to be described were standardized, however, descriptions might become stereotyped, the Council has therefore decided to leave the existing free descriptions unchanged.

Nevertheless, because information about research and development activities is important, for investors to assess managerial policies and future profitability, such information is encouraged to disclose voluntarily and positively.

V. Effective Date

The Council has decided that the Accounting Standards for Research and development Costs are to be applied to financial statements for fiscal years commencing on or after April 1, 1999.

Having regard to the possible effects of applying these Standards, the Council has also decided to allow businesses to continue to apply the existing accounting standards in respect of research and development costs that were already capitalized as assets before the introduction of these Standards.

The Council has requested the Japanese Institute of Certified Public Accountants, in consultation with interested parties, to prepare and issue practical guidelines on how to put these Standards into effect in the preparation of financial statements.

Accounting Standards for Research and Development Costs

March 13, 1998
Business Accounting Council

Section I. Definitions

1. Research and development "Research" is planned study and investigation aimed at discovering new knowledge. "Development" is materialization of research findings, and other knowledge, in the plan or design of a new product, service or production method ("products"), or in the plan or design for a remarkable improvement in an existing product.
2. Software "Software" is the generic name for programs in which instructions to run computers are combined.

Section II. Costs Included within Research and Development Costs

Research and development costs include all costs incurred for research and development, such as personnel expenses, material costs, depreciation charges on fixed assets, and indirect expenses allocated to research and development costs. (Note 1)

Section III. Accounting for Research and Development Costs

All research and development costs are to be charged to expenses, as incurred. Any portion of software production costs that corresponds to research and development costs is also to be charged to expenses, as research and development costs. (Note 2) (Note 3)

Section IV. Accounting for Software production Costs Not Included within Research and Development Costs

1. Accounting for custom-made software
   Production costs for custom-made software are to be accounted for in the same way as contracted construction.
2. Accounting for software for sale on the market
   The production costs of a master version of software (hereafter referred to as a "master") for sale on the market are to be capitalize as assets, expect for production costs that qualify as research and development costs. However, expenses paid to maintain the function of a master are not be capitalized as assets.

3. Accounting for software for internal use

Where any contract has been entered into to provide services, using software, to a specific outside party, and it is probable that profits will be realized from providing the services, the production costs of the software, as determined by the appropriate cost accumulation method, are to be capitalized as assets.

Where completed software has been purchased for internal use, and it is probable that profits will be realized, or costs reduced, through the internal use of the software, the costs of acquiring the software are to be capitalized as assets.

The costs of software installed into machinery or equipment are to be capitalized as assets, as a part of the costs of the respective machinery or equipment.

4. Classification of software

The costs of software for sale on the market and costs of software for internal use, that have been capitalized, are to be accounted for as costs of intangible fixed assets. (Note 4)

5. Depreciation of software

The acquisition costs of software included in intangible fixed assets are to be depreciated, by a method based on the quantity of expected sales, or any other method that is reasonable, taking based on the nature of the software.

The annual depreciation charge for any one accounting period, however, is not to be less than the average annual depreciation charge, over the remaining life of the software. (Note 5)

Section V. Notes to Application

The total amount of research and development costs included in general administrative expenses, and in production costs, for the current period is to be disclosed in notes to the financial statements. (Note 6)

Section VI. Scope of Application

1. Servicing contracts

These Standards are to be applied to any research and development activities carried out by outside parties on its behalf, under servicing contracts; but are not be applied to research and development that is carried out on behalf of outside parties.

Development of mineral resources

These Standards shall not be applied to activities specific to development of resources in the mining industry, such as exploration and drilling.

Interpretive Notes to Accounting Standards for Research and Development Costs

March 13, 1998
Business Accounting Council

Note 1: Costs included within research and development costs

   The costs of acquiring machinery, or any patent right, that is utilized only for a specific research and development project, and is not available for use for any other purpose, are to be charged to research and development costs, as incurred.

Note 2: Accounting for research and development costs

   Research and development costs may be charged either to general administrative expenses, or to production costs for the current period.

Note 3: Research and development costs in software production

   Where costs are incurred in developing software for sale on the market, the costs incurred until the completion of the master, and expenses paid to achieve any remarkable improvement in a master, or in any purchased software, are to be charged to research and development costs.

Note 4: Classification of software under production

   The costs of software under production are to be accounted for in a suspense account, under intangible fixed assets.

Note 5: Method to software under production

   Regardless of which method of depreciation is used, it is necessary to review the reliability of sales forecasts for every period, and write off the acquisition costs of the software, as expenses or losses, to the extent that the quantity expected to be sold decreases.

Note 6: Notes on research and development costs related to software

   Research and development costs related to software are to be disclosed in notes to the financial statements, as a component of research and development costs.

C

Mathematical Appendix

T. Takemura

In this appendix, we draft a method for solving a dynamic optimization problem that each firm faces from $t = 0$ to $t = \infty$, as seen in Chaps. 8 and 9. Refer to Wildasin (1984) for the details concerning theory of Tobin's q with multiple capital goods.

We yield the following dynamic optimization problem:

$$\max_{\{N(t),I(t)\}_{i=0}^{\infty}} \int_0^{\infty} \Big[F(K(t), N(t)) - \sum_{j=1}^{J} I_j(t) - \sum_{l=1}^{L} N_l(t) \Big] \mu(t) dt \quad \text{(C.1)}$$

$$\text{subject to } \dot{K}_j(t) = I_j(t) - \delta_j K_j(t) \text{ for all } j = 1, \ldots, J \quad \text{(C.2)}$$

where $F(\cdot, \cdot)$, $\Gamma(\cdot, \cdot)$ and $\mu(\cdot)$ are a production function, an organizational adjustment cost function, and a discount function, respectively. Specifically, each firm makes capital investments in several different asset types $I_j(t)$ with the vector of investment denoted by $I(t) = (I_1(t), \ldots, I_J(t))$, and expenditures in some variable costs $N_l(t)$ with the vector of variable inputs denoted by $N(t) = (N_1(t), \ldots, N_L(t))$, with the goal of maximizing the market value of the firm. For each $j = 1, \ldots, J$, $\delta_j > 0$ represents the (not necessarily time-invariant) proportional rate of depreciation for each capital good.

We assume that production function $F : K \times N \to \Re$ is linear homogeneous and concave, and organizational adjustment cost function $\Gamma : K \times I \to \Re$ is linear homogeneous and convex.[1] In addition, we assume that the production function satisfies the Inada condition.[2] The Inada condition states that firms cannot produce a good or service if they have no capital goods. In other words, this condition implies that capital goods are essential for production.

The discount function $\mu(t)$ is positive and decreasing with time t. That is, $\mu(t) > 0$ and $\mu'(t) < 0$. Because we obtain the present value of future cash

---

[1] We assume that these are real value functions.

[2] A function $f(\cdot)$ satisfies Inada condition if $f(0) = 0$, $f'(0) = \infty$, and $f'(\infty) = 0$. Inada condition means that a capital is absolutely essential for production.

flows of the firm and the solution that we seek converges, this condition is needed.

From Eqs. C.1 and C.2, we can write the Hamiltonian, H, as the following.

$$H = \left[F(K(t), N(t)) - \sum_{j=1}^{J} I_j(t) - \sum_{l=1}^{L} N_l(t)\right]\mu(t)$$

$$+ \sum_{j=1}^{J} \lambda_j(t)[I_j(t) - \delta_j K_j(t)] \tag{C.3}$$

where for every $j = 1, \ldots, J$, λ_j is costate variable.

At first, we assume that this problem has optimal solutions, $\{I^*(t)\}_{t=0}^{\infty}$ and $\{N^*(t)\}_{t=0}^{\infty}$. Halkin (1974) gives a proof for the existence of a solution. Then, this optimal solution should satisfy the following conditions; for all $t = 1, 2, \ldots$,

$$F_{N_l} - 1 = 0 \tag{C.4}$$

$$(\Gamma_{I_j} + 1)\mu = \lambda_j \tag{C.5}$$

$$\dot{\lambda}_j = \lambda_j \delta_j - (F_{K_j} - \Gamma_{K_j})\mu \tag{C.6}$$

for any $j = 1, \ldots, J$ and $l = 1, \ldots, L$. Furthermore, the optimal solution should satisfy the following condition: for all $j = 1, \ldots, J$,

$$\lim_{t \to \infty} \lambda_j(t)K_j(t) = 0 \tag{C.7}$$

which is called the transversality condition.

Note that by applying the Mangasarian theorem, and with the properties of F and Γ, the sufficient condition for the optimal solution is also satisfied.

Hayashi (1982) proves that in the case of single capital goods

$$V(0) = \lambda(0)K(0) \tag{C.8}$$

where $V(0)$ is the initial market value of the firm, determined by finding a relationship between marginal q and average q. Under the stated condition, this relationship between marginal q and average q is not lost even if it is the case of multiple capital goods.[3] Thus, for any capital stock, we gain

$$\lambda_j(0)K_j(0) = \lambda_j(0)K_j(0) - \lim_{t \to \infty} \lambda_j(t)K_j(t)$$

$$= -\int_0^{\infty} \left[\dot{\lambda}_j(t)K_j(t) + \lambda_j(t)\dot{K}_j(t)\right]dt \tag{C.9}$$

Moreover, by applying Eqs. C.4, C.5, and C.6, we obtain

---

[3]These assumptions are already given in this appendix.

$$-\dot{\lambda}_j(t)K_j(t) - \lambda_j(t)\dot{K}_j(t) = \big[(F_K - \Gamma_K)K_j(t) - (\Gamma_I + 1)I_j(t)$$

$$+ \sum_{l=1}^{L}(F_N - 1)N_l(t)\big]\mu(t) =: W_j(t). \text{ (C.10)}$$

By integrating and summing over $j = 1, \ldots, J$ in Eq. C.9, we obtain

$$\sum_{j=1}^{J}\lambda_j(0)K_j(0) = \sum_{j=1}^{J}\int_0^{\infty} W_j(t)dt = \int_0^{\infty}\sum_{j=1}^{J} W_j(t)dt \qquad \text{(C.11)}$$

Finally, by applying the Eular theorem and the homogeneity of both F and Γ, we obtain the following relationship concerning Tobin's q.

$$\sum_{j=1}^{J}\lambda_j(0)K_j(0) = \int_0^{\infty}\big[F^* - \Gamma^* - \sum_{j=1}^{J}I_j^*(t) - \sum_{l=1}^{L}N_l^*(t)\big]\mu(t)dt \quad \text{(C.12)}$$

Equation C.12 represents the sum of initial capital goods weighted by multiple shadow price λ_j, for every $j = 1, \ldots, J$, that is, shade value is equal to the present market value of the firm. In Chap. 8 and Chap. 9, we specify Eq. C.12 and estimate Eqs. 8.7 and 9.1.

References

1. Baltagi BH (1995) Econometric analysis of panel data, Wiley, New York, pp1-257
2. Ban K (1989) Macro-econometric analysis: validity and assessment of model analysis [J]. Yuhikaku, Tokyo, Japan, pp57-67
3. Berndt ER, Malone TM (1995) Information technology and the productivity paradox: getting the questions right. Economics of Innovation and New Technology 3:177-182
4. Berndt ER, Morrison CJ (1991) Computers aren't pulling their weight. Computer-World, December 9, 23-25
5. Berndt ER, Morrison CJ (1995) High-tech capital formation and economic performance in US manufacturing industries: an exploratory analysis. Journal of Econometrics 65:9-43
6. Berndt ER, Morrison CJ, Rosenblum LS (1992) High-tech capital formation and labor composition in US manufacturing industries: an exploratory analysis. National Bureau of Economic Research Working Paper 4010:1-39
7. Black SE, Lynch LM (2000) What's driving the new economy: the benefits of workplace innovation. National Bureau of Economic Research Working Paper 7479:1-41
8. Board of Business Accounting (1998) Statement of opinions on the establishment of accounting standards for research and development costs [J]. Ministry of Finance, Tokyo, Japan, 1-8
9. Brand H, Duke J (1982) Productivity in commercial banking: computers spur the advance. Monthly Labor Review 105:19-27
10. Bresnahan T, Brynjolfsson E, Hitt L (1999) Information technology, workplace organization, and the demand for skilled labor: firm-level evidence. National Bureau of Economic Research Working Paper 7136:1-39
11. Bresnahan T, Brynjolfsson E, Hitt L (2002) Information technology, workplace organization, and the demand for skilled labor: firm-level evidence. Quarterly Journal of Economics 117:339-376
12. Bresnahan TF (1986) Measuring spillovers from technical advances: mainframe computers in financial services. American Economic Review 76:742-755
13. Brynjolfsson E (1996) The contribution of information technology to consumer welfare. Information Systems Research 7:281-300
14. Brynjolfsson E, Hitt L (1993) Is information systems spending productive? New evidence and new results. Proceedings of the 14th International Conference on Information Systems, Vancouvar, British Columbia, December 47-64

15. Brynjolfsson E, Hitt L (1995) Information technology as a factor of production: the role of differences among firms. Economics of Innovation and New Technology 3:183-199

16. Brynjolfsson E, Hitt L (1998) Information technology and organizational design: evidence from micro data. http://ebusiness. mit. edu/erik/ITOD. pdf, cited on

17. Brynjolfsson E, Hitt L (2000) Beyond computation: information technology, organizational transformation and business performance. Journal of Economic Perspectives 14:23-48

18. Brynjolfsson E, Hitt L, Yang S (2000) Intangible assets: how the interaction of computers and organizational structure affects stock market valuations. http://ebusiness. mit. edu/research/papers/, cited on

19. Brynjolfsson E, Hitt L, Yang S (2002) Intangible assets: how the interaction of computers and organizational structure affects stock market valuations, Brookings Papers on Economic Activity: Macroeconomics (1):137-199

20. Brynjolfsson E, Malone T, Gurbaxani V, Kambil A (1994) Does information technology lead to smaller firms? Management Science 40:1645-1662

21. Brynjolfsson E, Yang S (1996) Information technology and productivity: a review of the literature. Advances in Computers 43:179-214

22. Brynjolfsson E, Yang S (1997) The intangible costs and benefits of computer investments: evidence from the financial markets. Proceedings of the International Conference on Information Systems, Atlanta, Georgia, December, 1997, 1-52

23. Daiwa Bank History Committee (1979) Daiwa Bank's history of six decades [J]. Daiwa Bank, Osaka, Japan, p147

24. Daiwa Bank History Committee (1988) Daiwa Bank's history for seven decades [J]. Daiwa Bank, Osaka, Japan, pp150-151, and p360

25. Daiwa Bank History Committee (1999) Daiwa Bank's history for eight decades [J]. Daiwa Bank, Osaka, Japan, pp218-221

26. Department of Commerce (1998) The emerging digital economy. Department of Commerce, U. S. , Washington, p29

27. DiNardo JE, Pischke JS (1997) The returns to computer use revisited: have pencils changed the wage structure too? Quarterly Journal of Economics 112:291-303

28. Economic Planning Agency (2000) Economic survey of Japan. Printing Bureau, Ministry of Finance, Tokyo, Japan, pp152-204 and pp362-363

29. Editing Committee for Handbook of Bank Regulations (1990) Handbook of bank regulations 1990 [J]. Kinzai Institute for Financial Affairs, Tokyo, Japan, p238

30. Endo K (1995) Hanshin great earthquake: action diary of BOJ Kobe branch manager [J]. Nihon Shinyo Chosa, Tokyo, Japan, pp1-187

31. Financial Supervisory Agency (1999) Manual used to inspect financial institutions [J]. Kin Ken no. 177, Financial Supervisory Agency, Tokyo, Japan, pp1-143

32. First Subcommittee of Business Accounting Council (1993) Statement of opinions on accounting standards for lease transactions [J]. Ministry of Finance, Tokyo, Japan, 1-8

33. Gordon RJ (2000) Does the 'new economy' measure up to the great inventions of the past? National Bureau of Economic Research Working Paper 7833:1-69

34. Gordon RJ (2002) Technology and the performance in the American economy. National Bureau of Economic Research Working Paper 8771:1-57

35. Greene WH (1997) Econometric analysis (3rd edn). Prentice Hall, New York, pp283-338

36. Griliches Z (1981) Market value, R&D, and patents. Economic Letters 7:183-187
37. Halkin H (1974) Necessary conditions for optimal control problems with infinite horizons. Econometrica 42:267-272
38. Hall BH (1993) The stock market's valuation of R &D investment during the 1980s American Economic Review 84:1-12
39. Hanaoka S (1995) The role and personnel training of an information system department [J]. Union of Japanese Scientists and Engineers, Tokyo, Japan, pp57-76
40. Harada H, Okamoto S (2001) Did IT revolution save the United States?[J] Economics Toyo Keizai, Tokyo, Japan 4:112-121
41. Hayashi F (1982) Tobin's marginal Q and average Q: neoclassical interpretation. Econometrica 50:213-224
42. Hayashi F, T Inoue (1991) The relation between firm growth and Q with multiple capital goods: theory and evidence from panel data on Japanese firms. Econometrica 59:731-753
43. Hiromatsu T, M Kurita, M Kobayashi, et al. (2000) Information technology and productivity of added values. ITME Discussion Paper, University of Tokyo 37:1-13
44. Hiromatsu T, M Kurita, M Kobayashi, et al. (2001) An econometric analysis on information technology. ITME Discussion Paper, University of Tokyo 83:1-33
45. Hiromatsu T, M Kurita, N Tsubone, et al. (1998) Quantitative analysis on labor-saving effect of informatization. ITME Discussion Paper, University of Tokyo 4:1-19
46. Hitt L (1999) Information technology and firm boundaries: evidence from panel data. Information Systems Research 10:134-49
47. Hitt L, Brynjolfsson E (1996) Productivity, profit and consumer welfare: three different measures of information technology's value. MIS Quarterly 121-142
48. Hitt L, Brynjolfsson E (1997) Information technology and internal firm organization: an exploratory analysis. Journal of Management Information Systems 14(2):79-99
49. Ishihara M (2000) Labour demand, wage and technological change in the US [J]. Japanese Journal of Labour Studies 42:60-70
50. Ito M (1999) New information technology makes system structural reform possible [J]. Weekly Kinzai, Kinzai Institute for Financial Affairs, Tokyo, Japan, April 26, pp20-23
51. Ito S, Kotani Y (1999) Research and development costs and computer software for accounting and tax practice [J]. Zeimu Kenkyukai Suppankyoku, Tokyo, Japan, pp1-244
52. Ito Y (2001) IT revolution and activation of Japanese economy[J]. In:Japan Center for Economic Research , Fujitsu Research Institte(eds) Research Report, Fujitsu Research Institute, Tokyo, Japan 102:30-50
53. Iwamura M (1996) Introduction to electronic money [J]. Nihon Keizai Shimbun, Tokyo, Japan, p26
54. Iwasa Y (1990) "Informatization" of financial institutions [J]. In: Information Industry Research Group (ed) Informatization and the modern society, The Institute of Economic and Political Studies, Kansai University, Research Series vol.72, Osaka, Japan, pp104-220
55. Japanese Bankers Association (Zenginkyo) (2001) The banking system in Japan. Zenginkyo, Appendix IV, p25, and p34

56. Japanese Institute of Certified Public Accountants (1999) Practical guidelines on accounting standards for research and development costs and software production costs [J]. Japanese Institute of Certified Public Accountants, Tokyo, Japan, 1-19

57. Jorgenson DW, Fraumeni BM (1989) The accumulation of human and nonhuman capital, 1948-84, In: Lipsey RE , Helen ST (eds) The Measurement of Saving, Investment, and Wealth, NBER Studies in Income and Wealth vol.52, University of Chicago Press, Chicago, USA, pp227-282

58. Jorgenson DW, Fraumeni BM (1992a) The output of the education sector, In:Griliches Z (ed) Output Measurement in the Services Sector, NBER Studies in Income and Wealth Vol. 55, University of Chicago Press, Chicago, USA pp303-338

59. Jorgenson DW, Fraumeni BM (1992b) Investment in education and US economic growth. Scandinavia Journal of Economics 94:51-70

60. Jorgenson DW, Motohashi K (2003) Economic growth of Japan and the United States in the information age. RIETI Discussion Paper Series, Research Institute of Economy , Trade and Industry 03-E-015:1-30

61. Jorgenson DW, Stiroh KJ (1995) Computers and growth. Economics of Innovation and New Technology 3:295-316

62. Jorgenson DW, Stiroh KJ (1999) Information technology and growth. American Economic Review 89(2):109-115

63. Jorgenson DW, Stiroh KJ (2000a) Raising the speed limit: US economic growth in the information age. Brookings Papers on Economic Activity 1:125-235

64. Jorgenson DW, Stiroh KJ (2000b) US economic growth at the industry. American Economic Review 90(2):161-167

65. Kamien MI, Schwartz NL (1991) Dynamic optimization: the calculus of variations and optimal control in economics and management (2nd edn). North-Holland, New York, pp3-118

66. Kamiyama T (2002) A case study of the mechanism of the system trouble at Mizuho Bank: from both sides of the information system and the management [J]. Office Automation 23(4):7-11

67. Kanto Data Center (1992) Kanto Data Center annual report [J]. Kanto Data Center, Tokyo, Japan, pp1-25

68. Kaplan R, Norton D (1996) The balanced scorecard: translating strategy into action. Harvard Business School Press, Boston, pp1-322

69. Kasuya S (1989) Econometric analysis of cost structure of banking industry: econometric analysis on business category of efficiency, technology progress, and factor substitution. Monetary and Economic Studies 8(2):79-118

70. Katagata Z (1989) Electronic banking [J]. NEC Culture Center, Tokyo, Japan, p79

71. Kimura T (2001) New examination manual and internal audit [J]. Economic Legal Research Institute, Tokyo, Japan, pp181-216

72. Koda H (2000) Business model patent [J]. Nikkan Kogyo Shimbun, Tokyo, Japan, pp42-52

73. Krueger AB (1993) How computers have changed the wage structure: evidence from micro-data, 1989-1984. Quarterly Journal of Economics 108:33-60

74. Kumasaka Y, Minetaki K (2001) IT economy [J]. Nippon Hyoron-sha, Tokyo, Japan pp1-178

75. Kuriyama N (2002) Measure of economic effects of information investment [J]. Final report for Grant-in-Aid for Scientific Research(C)(2)Tohoku University, Miyagi, Japan pp1-172

76. Kwon MJ, Stoneman P (1995) The impact of technology adoption on firm productivity. Economics of Innovation and New Technology 3:219-233
77. Lau LJ, Tokutsu I (1992) The impact of computer technology on the aggregate productivity of the United States: an indirect approach. Mimeo, Stanford University
78. Lee B (1998) Information technologies and the productivities of Japanese industries[J]. JCER Economic Journal 37:114-141
79. Lichtenberg FR (1995) The output contributions of computer equipment and personnel: a firm-level analysis. Economics of Innovation and New Technology 3:201-217
80. Mátyás L, Sevestre P (1992) The econometrics of panel data: handbook of theory and applications. Kluwer, Dordrecht, pp3-94
81. Maddala GS (1992) Introduction to econometrics (2nd edn). Prentice Hall, New York, pp191-213
82. Maimbo H, Pervan G (2002) A model of IS/IT investment and organizational performance in the banking industry sector. Proceeding of the 9th European Conference on Information Technology Evolution, Paris, France, July, 2002 15-16
83. Matsudaira J (1997) The effect of computerization on macroeconomics [J]. FRI Review, Fujitsu Research Institute 1:23-39
84. Matsudaira J (1998) The return to information technology investments in Japan[J]. FRI Review, Fujitsu Research Institute 2:43-57
85. Matsumoto K (2001) Informatization of economy and economic effect of IT [J]. DBJ Discussion Series, Development Bank of Japan, 22-1:1-65
86. Matsumoto Y (2001) Outline of business model patent and measure and subject of financial institutions [J]. Japan Post Research Monthly, Japan Post, Tokyo, Japan, August, pp93-100
87. Matsuura K, Y Takezawa, K Toi (2001) Money Crises and Economic Agents[J]. Nihon Hyoronsha, Tokyo, Japan, pp131-186
88. Minetaki K (2001) The effects of IT innovation on the labor market in Japan [J]. Economic Review, Fujitsu Research Institute 5:39-60
89. Ministry of Health, Labor and Welfare (2001) Analysis of labor economy : IT revolution and employment. Ministry of Health[J], Labor and Welfare, Tokyo, Japan http://www. mhlw. go. jp/wp/hakusyo/roudou/01/index. html , cited on
90. Mitsui Marine Safety Service Department (1993) Examples of accidents owing to computer systems and safety control [J]. Mitsui Marine, Tokyo, Japan, p4
91. Miyamura K (2002) Future of Mizuho Bank: comparison with example of Wells Fargo [J]. Office Automation 23(4):1-6
92. Miyao M (2001) What will change with the arrival of IY Bank [J]? PHP, Tokyo, Japan, p85
93. Miyazaki K, Kitamuro K (2001) Ubiquitous learning [J]. Office Automation 42:81-84
94. Morrison CJ (1997) Assessing the productivity of information technology equipment in US manufacturing industries. Review of Economics and Statistics 79:471-481
95. Morrison CJ, Berndt ER (1991) Assessing the productivity of information technology equipment in US manufacturing industries. National Bureau of Economic Research Working Paper 3582:1-23

96. Motohashi K(2003) Firm level analysis of information network use and productivity in Japan. RIETI Discussion Paper Series, Research Institute of Economy, Trade and Industry 03-E-021:1-24

97. Murata A (1999) Birth of debit card mega-market [J]. NTT, Tokyo, Japan, pp31-36

98. Nakaizumi T (1998) Survey of econometric studies of Japanese-US information investment [J]. In: Econometric analysis of economic effects of IT. Center for Global Communications, International University of Japan, Kokumin Keizai Research Institute, pp1-30

99. Nakano A (2000) Broadband society comes [J]! PHP, Tokyo, Japan, p44, and p187

100. Nikkei Computer (2002) Why did system troubles occur? Lessons from Mizuho [J]. Nikkei BP, Tokyo, Japan, pp8-74

101. Nishi T (2000) A revolution in the digital appliances industry [J]. PHP, Tokyo, Japan, p51

102. Nishigaki T (2001) IT revolution [J]. Iwanami Shoten, Tokyo, Japan, p159

103. Nishikawa Y (2001) Price index of computer in Japan-US. Monthly Report of Institute for Posts and Telecommunications Policy, November, pp4-15

104. Nishimura M (2000) Information and financial technology: two definite factors for financial industry [J]. Diamond Harvard Business vol 25, p 5

105. Nose T (2002) Security and risk management: lessons from troubles of Mizuho Bank [J]. Office Automation 23 (special edition IV):12-16

106. Obara M, Otake F (2001) The impact of computer use on wage structure [J]. Japanese Journal of Labour Studies 494:16-30

107. Ogawa K, Kitasaka S (1998) Asset market and business cycle[J]. Nihon Keizai Shimbun, Tokyo Japan pp1-299

108. Ogura N, Shimazaki T (2001) Research about balance score card construction in financial industry [J]. Kaikei 159: 82-93

109. Oliner SD, Sichel DE (1994) Computers and output growth revisited: how big is the puzzle? Brookings Papers on Economic Activity 2:273-334

110. Oliner SD, Sichel DE (2000) The resurgence of growth in the late 1990s: is information technology the story? Journal of Economic Perspectives 14:3-22

111. Osaki S, Iimura S (2001) Internet banking: truth and falsehood of network finance [J]. Nihon Keizai Shimbun, Tokyo, Japan, pp205-218

112. Otake F (2001) IT effect to employment [J]. Economics 4:80-86

113. Otsubo T (2001) Establishing smart card specifications by Japanese Bankers Association [J]. Weekly Kinzai, Kinzai Institute for Financial Affairs, Tokyo, Japan, April 9, 18-23

114. Prasad B, Harker PT (1997) Examining the contribution of information technology toward productivity and profitability in US retail banking. Financial Institution Center Working Paper, The Wharton School, University of Pennsylvania 97-09:1-35

115. Research Bureau of Economic Planning Agency (2000) IT effect to productivity: confirm the possibility of new economy in Japan [J]. Report of policy effect analysis , Economic Planning Agency, Tokyo, Japan 4:1-19

116. Research Group of Financial IT (2000) Debit card revolution [J]. Takarajima, Tokyo, Japan, pp153-156

117. Roach SS (1991) Services under siege: the restructuring imperative. Harvard Business Review 39:82-92

118. Segawa S (1993) Science of cards [J]. Kodansha, Tokyo, Japan, p27
119. Shimizu Y (1998) IT and wage differential survey of US econometric studies [J]. Monthly Report of Institute for Posts and Telecommunications Policy 115:164-172
120. Shinjo K (2000) IT capital stock and economic growth. Journal of Business Administration, Kwansei Gakuin University 47:1-19
121. Shinjo K, Zhang X (1999) Investment in the IT capital:US-Japan comparison. Kokuminkeizaizashi, Kobe University 129(6):1-16
122. Shinjo K, Zhang X (2003) Productivity analysis of IT capital stock: The USA-Japan comparison Journal of the Japanese and International Economies 17:81-100
123. Shinozaki A (1996) Analysis of the primary causes and economic effects of information-related investment in the United States and trends in Japan. JDB Research Report 59, Development Bank of Japan, Tokyo, Japan 1-53
124. Shinozaki A (1998) An empirical analysis of information-related investment and its impact on Japanese economy. JDB Research Report no. 59-02, Development Bank of Japan, Tokyo, Japan 1-34
125. Shinozaki A (1999) The nature of information technology revolution [J]. Toyo Keizai Shinpo-Sha, Tokyo pp1-227
126. Shinozaki A (2001a) Economic impact of information technology innovation [J]. Nippon Hyoron-Sha, Tokyo pp1-296
127. Shinozaki A (2001b) Investment in information technology and related labor force: industry level analysis in Japan[J]. Keizaigakukenkyu, Kyushu University 68:219-235
128. Shiratsuka S(1998) Economic analysis of price. University of Tokyo Press pp1-257
129. Shukuwa J (2000) All about the clearing systems [J]. Toyo Keizai, Tokyo, Japan, p53, and p203
130. Sichel DE (1997) The computer revolution: an economic perspective. The Brookings Institution, Washington DC, pp1-152
131. Siegel DE (1997) The impact of computers on manufacturing productivity growth: a multiple-indicators, multiple-causes approach. Review of Economics and Statistics 79:68-78
132. Siegel DE, Griliches Z (1992) Purchased services, outsourcing, computers, and productivity in manufacturing. In: Griliches Z (ed) Output measurement in the service sectors. University of Chicago Press pp429-458
133. Solow RM (1957) Technical change and the aggregate production function. Review of Economic Statistics 39:312-320
134. Solow RM (1987) We'd better watch out. New York Times Book Review, July 12 p36
135. Steiner RL (1995) Caveat! Some unrecognized pitfalls in census economic data and the input-output accounts. Review of Industrial Organization 10:689-710
136. Sugimura M (1999) Innovation of financial delivery channels: coping with diversified customer needs [J]. Economic Legal Reseach Institute, Tokyo, Japan, pp86-118
137. Suruga T (1991) The effect of computerization on employment in Japanese banking [J]. Japanese Journal of Labour Studies 380:28-38
138. Takeda K. (1998) Deregulation and productivity change in banking industry: an empirical study of deposit rate deregulation in Japan[J]. ICER Economic Journal 37:19-57

139. Takemura T (2002) Analysis about computerization investment disclosure in the banks of Japan [J]. http://www. rcss. kansai-u. ac. jp/, cited on

140. Takemura T (2003) Information system investment , productivity and efficiency in Japanese banking industry: evidence at firm-level using a stochastic frontier approach[J]. RCSS Discussion Paper, Researcb Center of Socionetwork Strategies, Kansai University 11:1-33

141. Tanaka H (2001) Reaction of Japanese firms to IT revolution: firm-level analysis on spillover effect and productivity growth of IT related investment[J]. Proceedings of the Eastern Meeting of Japan Economic Policy Association, Chuo University, Tokyo January 27

142. The Center for Financial Industry Information Systems (1994) The white paper on financial information systems [J]. Zaikei Syouhou, Tokyo, Japan, p501

143. The Center for Financial Industry Information Systems (1996) The white paper on financial information systems [J]. Zaikei Syouhou, Tokyo, Japan, in press

144. The Center for Financial Industry Information Systems (1999) White paper on financial information systems [J]. Zaikei Syouhou, Tokyo, Japan, in press

145. The Center for Financial Industry Information Systems (2000a) The white paper on financial information systems [J]. Zaikei Shouhou, Tokyo, Japan, pp5-532

146. The Center for Financial Industry Information Systems (2000b) FISC information system audit guidelines for financial institutions [J]. Zaikei Shouhou, Tokyo, Japan, pp1-130

147. The Center for Financial Industry Information Systems (2004a) Financial information systems in Japan. The Center for Financial Industry Information Systems, Tokyo, Japan, p6

148. The Center for Financial Industry Information Systems (2004b) The white paper on financial information systems [J]. Zaikei Syouhou, Tokyo, Japan, p272

149. Tobin J (1969) A general equilibrium approach to monetary theory. Journal of Money, Credit and Banking 1:15-29

150. Togashi N (2000) Banks' revival with IT revolution [J]. Jiji press, Tokyo, Japan, pp57-63

151. Toshida S, Japan Center for Economic Research (2000) IT revolution of Japan [J]. Nihon Keizai Shimbun, Tokyo, Japan pp1-291

152. Tsuchiya T, Harada R, Tomaru J (1983) Modern banking accounts [J]. Taga Syuppan, Tokyo, Japan, p170, and p173

153. UFJ Institute (2000) Front line of IC card business [J]. Kogyo Chosakai, Tokyo, Japan, pp112-122

154. Ukai Y (1997) Cross section analyses of information system investment in the banking industry using JMP software, Version 3. 1 [J]. The 16th SAS User Conference Proceedings, Yebisu Garden Palace, Tokyo, Japan, September, 1997 231-332

155. Ukai Y (2002) Human resources: the key factor to information system in banking industry [J]. Nikkei shimbun, Tokyo, Japan, January 22, p 25

156. Ukai Y, Kitano M (2002) An estimation of production functions including computer related investment in Japanese banking industry [J]. Workshop on Information System Investment Discussion Paper, Kansai University 5:1-14

157. Ukai Y, Takemura T (2001) Panel data analysis of computer software assets in the banks of Japan: estimation from the financial reports [J]. Economic Review of Kansai University 51:333-351

158. Ukai Y, Watanabe S (2001) Information technology investment in Japanese banks: panel data analysis [J]. Economic Review of Kansai University, Kansai University 51:51-81

159. Ukai Y, Watanabe S (2004) Interdependency between information system investment and human capital or organization: post master's questionnaires analysis [J]. Economic Review of Kansai University 54:355-376

160. Wildasin DE (1984) The Q theory of investment with many capital goods. American Economic Review 74:203-210

161. Wilson DD (1995) IT investment and its productivity effects: an organizational sociologist's perspective on directions for future research. Economics of Innovation and New Technology 3:235-251

162. Workshop on Information and Communication Policy (2002) Analysis of IT effect to economy [J]. Ministry of Internal Affairs and Communications, Tokyo, Japan 1-65

163. Yamada B, Sekiguchi M (1989) The post-third-generation on-line system and the strategic information system for banks [J]. Kinzai Institute for Financial Affairs, Tokyo, Japan, pp6-9, and p31

164. Yamaguchi S (2001) Reliability and security [J]. In: Murata M (ed) Internet as social infrastructure, Iwanami Internet Series no. 6, Iwanami Shoten, Tokyo, Japan, pp81-127

Index

1992 London explosion, 17, 20
1998 R&D standards, 107, 108, 110, 114, 115, 118, 120, 123–125 , 167, 170, 174–176, 178–180, 182
7 days/24-h, 11, 31, 32, 41

accounting system, 129, 133–136, 140, 154
adjustment cost, 165, 166
ADSL (asymmetric digital subscriber line), 46
aged deterioration, 59
aggregate level, 127
agriculture cooperative, 187
ANSER (Automatic Answer Network System for Electronic Request), 16
Article 21 of the Banking Law, 127
ATM (automatic teller machine), 8, 31, 32, 37–39, 41, 42, 49, 84, 184
average q, 165, 218
average age, 138
average annual income, 138, 141

bad debt, 169, 182
balance sheet, 80
BANCS (Bank Cash Service), 13
Bank of Japan, 127, 143, 146, 148, 155
banks of the dependent type, 148
batch-processing system, 3
BEA (Bureau of Economic Analysis), 58, 65, 66, 155
BLS (Bureau of Labor Statistics), 65, 72

Brynjolfsson E, 57–59, 63, 72–74, 76–83, 85–87, 97, 98, 165, 166, 184
BSC (balanced score card), 125, 184, 185
bubble economy, 62, 79
business cycle, 55, 61

CAFIS (Credit and Finance Information System), 16
cash card, 42, 44–46
cash corner, 12
cash flow, 218
CAT (Credit Authorization Terminal), 16
CATV (cable television service), 46
CD (cash dispenser), 8, 84, 184
cellular phone, 31, 41, 42
CIF (customer information file), 9
CIO (Chief Information Officer), 101
click and mortar, 32, 34
CLS (continuous linked settlement), 30
CMS (cash management service), 15
Cobb-Douglas production function, 59, 62, 66, 67, 71–75, 77, 78, 82, 162
collection center, 4
complementary, 82
computer application industry, 60
computer equipment, 109, 110, 114, 117, 118, 120–123, 165, 166, 168, 169, 173, 175–183
computer hardware investment, 100
computer software, 92, 98–101, 103, 104
computer software asset, 92, 165, 166, 168, 169, 172, 173, 175, 177–183

computer software investment, 100, 166, 176, 179, 182
computer supply industry, 60
computerization, 84
consumer-surplus effect, 56
contribution of IT capital, 57, 60–63, 68, 69
contribution ratio of IT capital, 60–63, 69
convoyed system, 29
correspondence policy, 4
correspondent arrangement, 8
cost cutting, 56
credit association, 187
credit card, 45
Credit Information Network, 17
credit union, 187
cross-section, 127, 148
cross-tab, 83
customer center, 15
cyber-terrorism, 34

debit card, 45
decision making, 82
delivery channel, 30–32, 35, 39–42, 47
demand-creation effect, 56
deregulation, 29, 31, 36
development cost, 128
development time of the information system, 128
discount function, 217
Divisia index, 75
Domar weight, 66
DoS (denial of service), 47
dynamic cost function, 65
dynamics optimization, 78

e-Japan, 55, 56
EB (electronic banking), 10, 16, 32, 38
EC (electronic commerce), 32
economic growth, 55, 57–62, 64, 67–69
economic growth rate, 58–63, 68, 69
elasticity of production with respect to IT labor, 75
employment effect, 56
Endo K, 23
external economy effect, 59

factor growth rate, 59

factorial analysis, 84
factory automation, 61
finance theory, 183
financial deregulation, 129
financial lease transaction, 110, 122, 168, 169
financial statement, 92, 94, 95, 102, 105, 165, 167, 169, 172, 180, 182, 184, 185
Financial Statements Overview, 129, 130, 140, 143
firm banking, 32
firm value, 78
firm value approach, 71, 81
firm value effect, 56
FIRST (Funds Information Relay System), 17
first-generation on-line system, 5–7, 9–11, 16, 25, 94
fisher cooperative, 187
fixed capital matrix, 67
fixed effects model, 73, 158
front-end system, 10
FSA (Financial Services Agency), 43
FTTH (fiber to the home), 46

general capital equipment ratio, 62
great Hanshin earthquake, 17, 21
Griliches Z, 64, 65, 69
growth accounting, 58–60, 62, 64–66, 68, 71, 72, 76

Hanaoka S, 27
Harker PT, 93, 98
Hausman test, 159
Hayashi F, 79, 165, 218
Headquarters for Promotion of Advanced Information and Communications Society (IT Strategy Headquarters), 55
Herstatt risk, 30
Hitt L, 57–59, 63, 72–74, 76, 77, 80–83, 86, 87, 166, 184
home banking, 32
hot standby, 12
hub-and-spoke type branch, 37
hub-and-spoke type system, 14
human capital, 57, 81–83
human resource, 169

hybrid system, 30

i-mode, 42
IC card, 6, 31, 34, 44–46
Iimura S, 41
IMT-2000, 42, 47
in-branch store, 37
in-store branch, 37
independent strategy, 148
information administration policy, 34
information system asset, 94, 107, 165,
   169, 173–180, 182, 183
Information System Asset I, 86, 100,
   125, 157, 159, 162
Information System Asset II, 100, 125,
   157, 159, 169
Information System Asset III, 100
information system department, 94–96,
   98
information system development cost,
   144–148
information system investment, 80,
   91–93, 96, 99–101, 104, 127–129,
   146, 148, 165, 166, 176, 179, 180,
   184, 185
Information System Investment I, 100,
   128, 153, 154, 158
Information System Investment II, 100,
   158
Information System Investment III, 100
information system management, 95
input-output analysis, 57
input-output table, 67–69
intangible asset, 76, 80
integrated system for ordinary accounts
   and term deposits (time deposits)
   (sogo koza), 8
international specialization hypothesis ,
   58
Internet banking, 40, 41
Inverse Auction, 50
investment standard, 129
ISO (International Organization for
   Standardization), 45
IT (information technology), 183
IT accounting, 125
IT equipment ratio, 62
IT revolution, 70
Ito M, 14

Iwamura M, 93, 96, 97
Iwasa Y, 8, 9

Japanese Bankers Association, 4, 19
Japanese classical abacus (soroban), 3
JIS I type (International Organization
   for Standardization); ISO), 45
JIS II type (Japanese Industrial
   Standards), 45
joint or cooperative branch, 37
Jorgenson DW, 58–61, 64, 66, 68, 69, 97

Kamiyama T, 44
Kaplan R, 184
Katagata Z, 12
keiretsu subsidiary, 135–140, 142
key-punchers, 3
Kimura T, 52
Kitano M, 98
Koda H, 50
Kuriyama N, 57, 64, 67, 69, 72, 82, 87

labor productivity, 59, 83
legal disclosure, 111, 112, 124
Lichtenberg FR, 72, 74, 76, 77, 86
loan and bills discounted, 156, 157, 159,
   169, 170, 174–176, 178–182
long learning lags hypothesis, 57

management strategy effect, 56, 84
marginal q, 165, 218
marginal product, 59
marginal productivity, 76, 78
market value approach, 149
Matsumoto Y, 50
mega bank, 42, 43, 50
message broker, 14
MICR (magnetic ink character
   recognition), 4
micro data, 98, 104, 127, 143
MICS (Multi Integrated Cash Service),
   13, 17
Ministry of Economy, Trade and
   Industry, 83
Ministry of Finance, 29, 36, 43
mismanagement hypothesis, 58
mismeasurement hypothesis, 57
Miyamura K, 44
Miyao M, 39

Miyazaki K, 47
mobile banking, 41
mobile branch, 37
Motohashi K, 58, 62, 69
Multi-Payment Network, 30
Murata A, 29
mutually complementary relationship,
    81

Nagaoka H, 93, 96
Nakano A, 34, 46
narrowly-defined information system,
    129, 134–137, 154
nationwide bank, 5, 7, 13, 15, 18, 28,
    29, 91, 94–96, 129, 130, 136–138,
    143, 173, 176, 177, 180, 187
nationwide mega bank, 103, 104, 136,
    137, 139, 189
Nichigin Net (Bank of Japan Financial
    Network System), 16, 17, 187
NIPAs (National Income and Product
    Accounts), 58
Nippon Telegraph and Telephone Public
    Corporation, 8, 18
Nishi T, 47
Nishigaki T, 34
non-keiretsu, 135, 138, 139
Norton D, 184
Nose T, 44

off-line system, 3–5, 8, 24
office automation, 61, 84
office machinery, 168, 169
offsetting factors hypothesis, 57
Ogura N, 184
OJT (on-the-job training), 80
Oliner SD, 57–61, 66, 68
on-line system, 3, 5–8, 12, 14, 15, 17–19,
    24, 25, 109, 171
One-Click Method, 50
one-to-one marketing, 32
operating cost, 83
operation starting year, 130, 131, 134
operational lease transaction, 110
operational risk, 30, 34, 47
organization change, 169
organization effect, 81
organization reform effect, 56, 81
organization scale effect, 81

organization structure, 184
organizational adjustment cost function,
    166, 217
Osaki S, 41
Otsubo T, 46
outsourcing, 6, 15, 25–27, 101, 128, 129,
    137–139, 142, 148
outsourcing cost, 80

package software, 83
panel data, 165–167
PCS (punched card system), 3, 25
personnel cost, 92, 94, 99, 100, 103, 104,
    141
portfolio selection, 183
positive productivity, 56
post master, 83
post-third-generation on-line system, 6,
    9, 14, 78, 166
Prasad B, 93, 98
present market value of the firm, 219
principal component analysis, 80
production function approach, 71, 149
productivity effect, 57
productivity paradox, 56–58, 64–66, 69
profit maximization, 79
proxy variable, 83
public telecommunications law, 8

quick corner, 12
quick lobby, 12

random effects model, 158
rank correlation coefficient, 83
real-time gross settlement, 30
redistribution hypothesis, 58
regional bank, 91, 129, 130, 134, 135,
    138, 139, 141, 143, 173, 176,
    178–180, 187, 189
renewal time, 129
required development time, 134, 135
restructuring cost, 80
risk management, 30
Roach SS, 64
robot-retailing branch, 37
ROI (return on investment), 183

sample selection bias, 82
second regional bank, 129, 130, 143, 187

second-generation on-line system, 5, 6, 9–12, 16, 25, 100, 171
securities operation and management system, 154
Segawa S, 45
self-management team, 81
Setagaya cable fire, 10, 17–19
shadow price, 167, 219
Shiba K, 95–97
Shimazaki T, 184
Shinozaki A, 57, 58, 62–64, 66–69
Shukuwa J, 30
Sichel DE, 57–61, 66, 68, 93, 97
Siegel DE, 57, 64, 65, 69
software engineer, 102
Solow R, 57
Spearman partial rank correlation coefficient, 82
start time of the information system, 128
Stiroh KJ, 58–61, 64, 66, 68
stochastic frontier model, 78
stock market value model, 166
Suda K, 95, 97, 98, 168
Sugimura M, 36, 37
SWIFT (Society for Worldwide Interbank Financial Telecommunication), 16
system auditing, 31, 44, 50
system department, 6, 14, 24–27
system development cost, 80
system development section, 84
system operation, 139

Takeda K, 96–98
Takemura T, 72, 77, 78, 80, 86, 99, 113, 124, 166
Teikoku Databank, 82
telemarketing, 40
telephone banking, 32, 39, 40
teletypewriter, 4
TFP (total factor productivity), 57–62, 65–67

the Center for Financial Industry Information Systems, 6, 13, 14, 16, 33, 37, 42, 44, 45, 47, 51, 91, 94, 96
theory of Tobin's q, 183, 217
third-generation on-line system, 5, 6, 11–14, 26, 28, 94, 96, 100, 137, 171
time-lag effect, 76
Tobin's q, 165, 166, 182, 183, 219
Togashi N, 14
total (integrated) on-line system, 7
total asset, 129, 130, 144–148
total market value, 168, 176, 179, 180, 182, 183
trans-log cost function, 62
trans-log production function, 71, 73
transversality condition, 218

ubiquitous banking, 47
Ukai Y, 72, 77, 80, 82–84, 86, 91–99, 146, 166
unmeasured correlated intangible asset, 165, 167
user cost, 59, 60, 62, 79

VIF (variance inflation factor), 157, 173
voluntary disclosure, 111, 112, 124

Watanabe S, 72, 80, 82, 83, 86, 91–94, 96–99
WISI (Workshop of Information System Investment), 91, 93–99
workshop innovation, 81

Y2K issue, 129
Y2K problem (Year 2000 problem), 15, 17, 23, 24
Yamaguchi S, 47, 99
Yang S, 59, 72, 76, 78–80, 86, 87, 165, 166, 184

Zengin system (Domestic Funds Transfer System), 8, 16, 22, 187